Letting in
THE DOG

W9-DEU-461

Opening hearts and minds
to a deeper understanding

Pat Blocker CPDT-KA

Hubble & Hattie

The Hubble & Hattie imprint was launched in 2009 and is named in memory of two very special Westies owned by Veloce's proprietors. Since the first book, many more have been added to the list, all with the same underlying objective: to be of real benefit to the species they cover, at the same time promoting compassion, understanding and respect between all animals (including human ones!)

Hubble & Hattie is the home of a range of books that cover all-things animal, produced to the same high quality of content and presentation as our motoring books, and offering the same great value for money.

More great Hubble & Hattie books!
Among the Wolves: Memoirs of a wolf handler (Shelbourne)
Animal Grief: How animals mourn (Alderton)
Babies, kids and dogs – creating a safe and harmonious relationship (Fallon & Davenport)
Because this is our home ... the story of a cat's progress (Bowes)
Bonds – Capturing the special relationship that dogs share with their people (Cukuraite & Pais)
Camper vans, ex-pats & Spanish Hounds: from road trip to rescue – the strays of Spain (Coates & Morris)
Canine aggression – how kindness and compassion saved Calgacus (McLennan)
Cat and Dog Health, The Complete Book of (Hansen)
Cat Speak: recognising & understanding behaviour (Rauth-Widmann)
Charlie – The dog who came in from the wild (Tenzin-Dolma)
Clever dog! Life lessons from the world's most successful animal (O'Meara)
Complete Dog Massage Manual, The – Gentle Dog Care (Robertson)
Confessions of a veterinary nurse: paws, claws and puppy dog tails (Ison)
Detector Dog – A Talking Dogs Scentwork Manual (Mackinnon)
Dieting with my dog: one busy life, two full figures ... and unconditional love (Frezon)
Dinner with Rover: delicious, nutritious meals for you and your dog to share (Paton-Ayre)
Dog Cookies: healthy, allergen-free treat recipes for your dog (Schöps)
Dog-friendly gardening: creating a safe haven for you and your dog (Bush)
Dog Games – stimulating play to entertain your dog and you (Blenski)
Dog Relax – relaxed dogs, relaxed owners (Pilguj)
Dog Speak: recognising & understanding behaviour (Blenski)
Dogs just wanna have Fun! Picture this: dogs at play (Murphy)
Dogs on Wheels: travelling with your canine companion (Mort)
Emergency First Aid for dogs: at home and away Revised Edition (Bucksch)
Exercising your puppy: a gentle & natural approach – Gentle Dog Care (Robertson & Pope)
For the love of Scout: promises to a small dog (Ison)
Fun and Games for Cats (Seidl)
Gods, ghosts, and black dogs – the fascinating folklore and mythology of dogs (Coren)
Helping minds meet – skills for a better life with your dog (Zulch & Mills)
Home alone – and happy! Essential life skills for preventing separation anxiety in dogs and puppies (Mallatratt)
Know Your Dog – The guide to a beautiful relationship (Birmelin)
Letting in the dog: opening hearts and minds to a deeper understanding (Blocker)
Life skills for puppies – laying the foundation for a loving, lasting relationship (Zuch & Mills)
Lily: One in a million! A miracle of survival (Hamilton)
Living with an Older Dog – Gentle Dog Care (Alderton & Hall)
Miaow! Cats really are nicer than people! (Moore)
Mike&Scrabble – A guide to training your new Human (Dicks & Scrabble)

Mike&Scrabble Too – Further tips on training your Human (Dicks & Scrabble)
My cat has arthritis – but lives life to the full! (Carrick)
My dog has arthritis – but lives life to the full! (Carrick)
My dog has cruciate ligament injury – but lives life to the full! (Haüsler & Friedrich)
My dog has epilepsy – but lives life to the full! (Carrick)
My dog has hip dysplasia – but lives life to the full! (Haüsler & Friedrich)
My dog is blind – but lives life to the full! (Horsky)
My dog is deaf – but lives life to the full! (Willms)
My Dog, my Friend: heart-warming tales of canine companionship from celebrities and other extraordinary people (Gordon)
Office dogs: The Manual (Rousseau)
One Minute Cat Manager: sixty seconds to feline Shangri-la (Young)
Ollie and Nina and ... daft doggy doings! (Sullivan)
No walks? No worries! Maintaining wellbeing in dogs on restricted exercise (Ryan & Zulch)
Partners – Everyday working dogs being heroes every day (Walton)
Puppy called Wolfie – a passion for free will teaching (Gregory)
Smellorama – nose games for dogs (Theby)
Supposedly enlightened person's guide to raising a dog (Young & Tenzin-Dolma)
Swim to recovery: canine hydrotherapy healing – Gentle Dog Care (Wong)
Tale of two horses – a passion for free will teaching (Gregory)
Tara – the terrier who sailed around the world (Forrester)
Truth about Wolves and Dogs, The: dispelling the myths of dog training (Shelbourne)
Unleashing the healing power of animals: True stories about therapy animals – and what they do for us (Preece-Kelly)
Waggy Tails & Wheelchairs (Epp)
Walking the dog: motorway walks for drivers & dogs revised edition (Rees)
When man meets dog – what a difference a dog makes (Blazina)
Wildlife photography – saving my life one frame at a time (Williams)
Winston ... the dog who changed my life (Klute)
Wonderful walks from dog-friendly campsites throughout the UK (Chelmicka)
Worzel Wooface: For the love of Worzel (Pickles)
Worzel Wooface: The quite very actual adventures of (Pickles)
Worzel Wooface: The quite very actual Terribibble Twos (Pickles)
Worzel Wooface: Three quite very actual cheers for (Pickles)
You and Your Border Terrier – The Essential Guide (Alderton)
You and Your Cockapoo – The Essential Guide (Alderton)
Your dog and you – understanding the canine psyche (Garratt)

Hubble & Hattie Kids!
Fierce Grey Mouse (Bourgonje)
Indigo Warrios: The Adventure Begins! (Moore)
Lucky, Lucky Leaf, The: A Horace & Nim story (Bourgonje & Hoskins)
Little house that didn't have a home, The (Sullivan & Burke)
Lily and the Little Lost Doggie, The Adventures of (Hamilton)
Wandering Wildebeest, The (Coleman & Slater)
Worzel goes for a walk! Will you come too? (Pickles & Bourgonje)
Worzel says hello! Will you be my friend? (Pickles & Bourgonje)

WWW.HUBBLEANDHATTIE.COM

First published in July 2019 by Veloce Publishing Limited, Veloce House, Parkway Farm Business Park, Middle Farm Way, Poundbury, Dorchester, Dorset, DT1 3AR, England. Tel 01305 260068/fax 01305 250479/e-mail info@hubbleandhattie.com/web www. hubbleandhattie.com. ISBN: 978-1-787113-05-3 UPC: 6-36847-01305-9. © Pat Blocker & Veloce Publishing Ltd 2019. All rights reserved. With the exception of quoting brief passages for the purpose of review, no part of this publication may be recorded, reproduced or transmitted by any means, including photocopying, without the written permission of Veloce Publishing Ltd. Throughout this book logos, model names and designations, etc, have been used for the purposes of identification, illustration and decoration. Such names are the property of the trademark holder as this is not an official publication. Readers with ideas for books about animals, or animal-related topics, are invited to write to the editorial director of Veloce Publishing at the above address. British Library Cataloguing in Publication Data - A catalogue record for this book is available from the British Library. Typesetting, design and page make-up all by Veloce Publishing Ltd on Apple Mac. Printed and bound in India by Replika Press PTY

Contents

Dedicated to dogs

ACKNOWLEDGMENTS

I cannot give adequate thanks to the following people who have rallied round to make this book the best it can be. Your assistance helps me to help more people and their dogs, and I am eternally grateful.

I'd like to hand a heartfelt thank you to Barrie Finger for reading the manuscript for content. You have both the eye of a mindful dog trainer and a friend who will give gentle, constructive criticism. You know me well enough to understand what I meant to say when I wasn't being clear. It's invaluable to have the assistance of a dear friend who's been on a similar professional and spiritual path.

Mark Holly of Mark Holly Studios deserves huge thanks for his ability to get into my head and turn my ideas into the glorious illustrations you see in this book. You were brave in agreeing to work with this former graphic designer. Two artists on one project could have gotten quite messy, but you handled it with typical humor and grace.

I am grateful to Michael Curran for so accurately capturing the spirit of Mr Mojo in the book's cover photo.

Thank you Patrina Odette for reading my manuscript with the eyes of a savvy dog mom and the heart of a supportive friend. You and your dogs backed me through some bumpy times, giving me a good laugh and steadfast encouragement.

Thank you Radleigh Valentine for gently nudging me into the world of animal communication. Your spiritual teaching and friendship have helped me to set foot on this path and to stay the course. Thank you for believing in me. You make me feel like Dorothy in *The Wizard of Oz* when the good witch told her that she always had the power. She just had to learn it for herself.

Pat Blocker
Colorado, USA

FOREWORD

When Pat Blocker asked me to write the foreword for her latest book, my left paw went up and I spun in place three times.

Well, not really. But those things would've happened if I were my dog. And that's my point! After all these decades of loving, adoring, and holding my 'four-pawed wonders' as dear as any human child, Pat Blocker has finally taught me how to speak Dog.

Let me ask you something ... How many times have you looked into the unconditionally loving eyes of your pooch and said "I just wish you could tell me how you feel?" For me that number is far too high to count.

This has been especially challenging for me as both my dogs are rescues. Their histories are mysteries. When they exhibit anxious behavior I used to have no idea if it was their unknown past influencing them or something in the here and now. This caused me great sadness and worry. I found myself constantly wishing I knew what they were trying to tell me.

Well, not any more! By writing *Letting in the dog*, Pat Blocker has applied her decades of experience to crack the canine code, and make speaking Dog simple for both you and me.

While I firmly believe that every puppy comes into this world instinctively knowing the two words for 'treat' and 'walk,' it turns out that the language of Dog is actually totally different from what most people imagine it is. Naturally, you'd think that, as a world-renowned psychic, this point would've been obvious to me. But for years I have just chatted away to my dogs, presuming that they were slowly figuring out what my words meant.

And, to an extent, they were. Our canine companions are quite brilliant – and yes – they do figure out certain words. But the truth is that they communicate in a completely different way. They understand us far more than we understand them.

But it doesn't have to be that way!

Letting in the dog is the super secret codebook to the language of Dog that you've always dreamed of having. Brilliant, funny, charming, and heart-opening, Pat Blocker has written the universal translator for having an open and loving dialogue with your best friend.

I started this foreword by saying that Ms Blocker's request caused me to raise my left paw and spin in place three times. Well, that's an inside dog joke, but you can be in on the joke as well! By reading this book you will discover that those two behaviors represent excitement but also nervousness. With Ms Blocker's guidance, you will learn how to 'listen to your eyes and hear what you see,' because that's how dogs communicate.

Of course, this is hardly the first book on understanding canine behavior. But what makes *Letting in the dog* different from those other books is that it represents the magical space where the science of dog training and the spirituality of animal communication come together.

Full of heart, humor, and wisdom, *Letting in the dog* is an 'honest to dog' page-turner for anyone wanting a closer relationship with their pooch. The connection I have with my 'fur-babies' has been expanded beyond my wildest dreams just by reading this book. I now look at my beloved pets and smile to myself because each action they display now means something. They are talking to each other, and because of this book, I am now in on the conversation.

Would you like to be in on the conversation, too? No problem. The ABC of the Language of Dog is right here in front of your eyes.

Courtesy Patrina Odette

Radleigh Valentine
Spiritual teacher, psychic, and bestselling Hay House author of *Angel Tarot Cards, Animal Tarot Cards,* & *How to Be Your Own Genie*

Introduction

This book is about communication and understanding between humans and those amazing creatures we call man's best friend. I've written about canine body language before, but this book is different. Here, my intent is to bridge the space between the academics of dog training and intuitive communication with dogs. We'll meet at the intersection of the science and the psychic. We'll gently lean in to the spiritual side of communicating with canines. I wish to present a palatable introduction for those skeptical of animal communication, and to create an agreeable starting point where strict academics can open up to the heart of training.

While we can have only a human understanding of the world, dogs can have only a canine understanding of it. Communication is the intersection of understanding: understanding each other. Reading dog body language and intuitively reading a dog are like two sides of the same coin. Body language must be taken in context, as context does as much to convey the meaning as does the gesture.

Is it metaphysical or mechanical? I suspect the answer to that question is, 'both.' But what's really important is that we ask the question about how to communicate with dogs. Then we should listen as much, if not more, than we speak. In doing so, we learn that to connect with a dog allows us to connect with ourselves and with our spirits.

In January of 2017 my friend, Radleigh Valentine, professional intuitive, speaker, author, radio host, and spiritual advisor, asked me to be a guest on his radio show *Magical Things*. Before he became my friend Radleigh was a client, so he knew fully how I operated as a dog trainer. He said he'd been thinking about having me as a guest on his show for some time, but there was a problem: he hadn't found the right intersection between our careers. His listeners would expect an animal communicator; someone who communicates with animals psychically. He posed that I'm not an animal communicator in the psychic sense, yet I do communicate with dogs, and do so with both a scientific approach and a touch of the intuitive. *This* was where our fields intersected.

That *Magical Things* radio experience was the inspiration for this book. *Letting in the Dog* is about more than reading dog body language. It's about finding that juncture between the academic and the spiritual: the joining of hearts and minds. It's about using both heart speak and mind speak – not just talk.

I've entitled this book *Letting in the Dog* because we must let dogs into our hearts as well as into our minds in order to have a real conversation. Often, dogs sit on the other side of the door waiting patiently and asking to come in. We cannot hear them speaking to us until we open the door and let them in. This book may not be the key to unlocking the secrets of human, canine, and planetary transfiguration, but my hope is that it helps you to open the door of your heart and to let in the dog. Ultimately, I am writing this book to help dogs and their people: to shine a light on the path of understanding, trust, and building stronger relationships through communication. Thank you for letting me share my journey. I hope you enjoy this walk with dogs as much as I do. Together we will make an amazing difference to the lives of dogs because they have always made an incredible difference to ours.

This is my personal story of connecting with dogs and how they have helped me to connect with myself. It is this connection that allows us to help dogs, who in turn help us. We are two spirits merging into one mind and one heart. This is how we heal. Many of us have lost touch with our intuitiveness and have come under the rule of our egos. We've become detached from our spirits and our true inner nature. In losing touch with our own inner nature, we lose our ability to relate intuitively to one another. Dogs can help us regain that understanding in order to relate to them, and to relate to ourselves and other humans.

There is no other species like the human species. There is no other species like the canine species. Nonetheless, when all is said and done, the similarities between humans and canines overwhelm the differences. Communication is where we find common ground. I hope that this book will make the world a kinder place for dogs and for their people: a place where we can coexist with compassion.

Courtesy Amy Faith Martin

Pat with Jett (left) and Penny Lane.

A NOTE TO READERS

You will notice the use of the pronouns 'he' and 'she' throughout this book when referring to our canine companions. The words are intended to be non-specific, and are not meant to show gender bias: this author has equal fondness for both male and female dogs.

This book uses the word 'owner' when referring to canine guardians or caretakers. I pondered at length about whether to use owner or guardian, or to call us 'pet parents.' Owner is used strictly in the interest of simplicity, and is not meant to imply that pets are dispensable property, or to devalue them as companions, family, and friends.

Enlighten up! Follow your intuition more than the rules

THERE IS A LONGING AMONG ALL PEOPLE AND CREATURES TO HAVE A SENSE OF PURPOSE AND WORTH. TO SATISFY THAT COMMON LONGING IN ALL OF US WE MUST RESPECT EACH OTHER.
Chief Dan George

Respect and truth are the underpinnings of any successful relationship. They demonstrate that we recognize the value of another being and build trust. Mindfulness of both sides of the conversation is paramount for understanding, no matter whether the conversation is with another human, one's self, or with our dog.

As a student of dog behavior and training, I've learned the rules and science of canine learning and behavior modification. As a student of life, I've learned a few lessons pertaining to spirit. I've been blessed with wise teachers of both human and non-human form, who've helped me to connect the dots between the academic and intuitive. In the foreword to my first book, *Taking the Lead without Jerking the Leash*, my friend, colleague, and fellow author Veronica Boutelle says, "Good dog trainers know their learning theory science and have excellent training mechanical skills. Great dog trainers also understand that training is as much about the human end of the leash as it is the canine end." I would add that a great trainer learns as much as she teaches, and listens as much as she speaks.

Effective training happens with good communication and through mindfulness. We must be present to win. When I say, 'win' I don't mean that training is a competition. Here, winning means a commitment to being nonjudgmental and to hold compassion in our hearts. A skillful communicator is a good listener and a perceptive observer. I have witnessed these qualities in top authorities on canine behavior. Many of these authorities are human, and many of them have four legs and fur.

Mindfulness in our relationships with dogs allows us to remove our egos, bust myths, drop assumptions, and open our minds to facilitate effective communication. Thus, we open our hearts to our dogs by being in the here and now. Relationships can be challenging, but overall, the desire is for them to be fun and loving. That desire exists on both ends of the leash. Two-way communication opens the door and breaks down barriers to having that fun and loving relationship.

Before I began writing this book and exploring the deeper, intuitive side of communicating with dogs, I'd been living in a less illuminated place. I've been on a spiritual path for many years, but I had a two-year period where I'd descended into fear and self-doubt. I compensated for that by pushing and punishing myself. Pushing and punishing myself didn't work; I should have known that. In dog training, I use positive reinforcement, which means I set dogs up for success and reward them for that success instead of punishing them for undesirable behaviors.

I should have practiced what I preach. I know that when animals are afraid or far outside their comfort zone, it's not a teachable moment. Optimum learning and communication take place in the space of calm. I needed to remember that as I overworked myself with self-imposed deadlines and pressure. The source of my worries was the state of humanity, politics, finances, and, well, you name it. I was taking on the troubles of the world, and in my mind it was up to me to fix them.

I had many conversations with a friend who is a spiritual teacher. He told me how I simply needed to shine my own light. My question was, 'How can being a dog trainer fix the world?' My friend shared a great analogy about life being similar to walking a maze. In the maze you're not sure where you are, but you know it's not where you want to be. As you're walking along, a giant rotten tomato drops right in your path. How unfair is that? You have places to go and limited time to get there. He told me to take the tomato as a signal that the path I'm on is not going to make me happy, and that I should choose another direction.

I said that I understood, but in my world that rotten tomato would be an enormous bag of dog poo, and I felt it was my responsibility to pick it up. I realized then and there that it's not my job to clean up after the entire world. I need to follow my heart on the path of my divine life purpose which, for me, is dog training. By shining my light, I can help others to shine theirs. Dogs help us to shine if we listen to them. Dogs help us to listen to ourselves as we begin to hear what they have to teach us. They help us hear our quiet inner voice over that of our ego, and they shine a light to guide us on our path.

Animals have a pure and unobstructed awareness that allows them to stay connected to the here and now. I believe that dogs have an innate 'knowing' of being needed and loved, and that their presence is all that's required for positive change. We humans forget that about ourselves, but, if we allow them to, dogs will help us find our way. If we are aware, we will find each other and shine together.

FINDING EACH OTHER: FOLLOWING OUR HEARTS AS A DOG FOLLOWS HER NOSE

We can study up on breeds. We can thoroughly assess our lifestyle and which type of dog is the best fit for us. We can research breeders, shelters and rescue groups. We can educate ourselves on every aspect of the care and feeding of a dog. And we should. I would never suggest getting a dog without the proper preparation. However, after all of the research, assessment, and education, the absolute final choice is made from the heart ... from both hearts. You'll just know you're right for each other when you meet.

It is in the universal scheme and, as they say, 'it was meant to be.' I believe that we don't necessarily get the dogs we want; we get the dogs we are meant to have. Each of my dogs has had his or her specific lesson to teach. Dogs come into our lives to teach us, and they leave our lives to teach us. Goodbyes are so very difficult. But I am grateful that each of my dogs profoundly enhanced my life and brought me unending joy. It's this very fact that makes it so difficult to say goodbye.

From the time I was a little girl I've loved animals. At a young age, I realized that the compassion I felt for them stemmed from the realization that they could not speak for themselves, and I wanted to speak for them. I became a dog trainer because I wanted to give dogs a voice. I want to help them be heard by helping people to better understand them. I now realize that animals actually can and do speak: we simply need to learn how to hear them. An open mind and open heart are of paramount importance in communicating with animals. If we don't listen, we don't hear and we don't learn.

This journey is about learning to communicate with dogs in order to forge a trusting and joyful relationship. It's not about simply training for improved obedience, or molding our dog's behavior to conveniently fit our lifestyle. The word 'improvement' implies that there is something wrong with the relationship. The notion of needing improvement keeps us in a state of wanting the relationship to be different, so, instead of this, let's first accept it and appreciate its present perfection. Only from here can we make changes for the better, if needed. By changing our perception we can learn what lessons our dogs have to teach *us* before we begin teaching them ours.

DISCOVER THEIR PERFECTION

I believe that each of our dogs (all pets, for that matter) enters our lives for a specific purpose. Every one of my dogs came into my life for a reason: to teach. I followed my heart each time I brought a new dog into my life, and, by doing so, I found my best teachers. And they are still teaching me – even from those who have passed over, I'm still discovering the perfection and the lessons of each dog's presence in my life. Here are their stories.

Finding Gypsy

Gypsy came into my life to teach me about responsibility.

Gypsy was the first dog I had as an adult. I'd grown up with dogs on our farm in Iowa, but these were farm dogs: part of the family, but also with jobs as gatekeepers and guardians of livestock.

I'm eternally grateful to Gypsy for educating me about the care and feeding of the canine species. Sadly, Gypsy died mysteriously at the young age of four, but her untimely demise taught me about the ephemeral quality and brevity of life.

Finding Jesse

Jesse came into my life to teach me about problem-solving.

Until Jesse, I had no idea of the power and cleverness that some dogs possess. Jesse knew how to feign an alert to danger, which would have her sister, Sadie, come running to the scene. Upon Sadie's arrival, Jesse would beeline it to the preferred sleeping spot that Sadie had just vacated. Similarly, Jesse would bark as if to say that she needed to go outside as a ruse to get Sadie to leave the bed. Not long ago, an animal communicator doing a reading with Sadie told me that Sadie knew Jesse's game and participated because it was fun. The joke was really on me the whole time.

Finding Sadie

Sadie filled a void in my life and came to teach me about laughter.

Sadie was my divorce dog. I brought her into my home after my husband and I parted ways. I was fully conscious that I was attempting to fill the void left by divorce, and that I was seeking unconditional love. Sadie was a long-coat German Shepherd. Funny; when I spoke to her breeder prior to my first visit, I'd asked if she was a long-coat because I didn't want the long hair. I didn't know that much about Shepherds at the time, and haven't a clue why I asked that question. But Sadie was awfully fluffy, and, well, I took her home anyway.

Sadie was a clown among clowns with a large personality to match her physical size. She was a big girl who looked even bigger because of that voluminous coat. Sadie indeed filled that post-divorce void, offered me unconditional love, and gifted me with laughter.

Finding Abbey

Abbey came into my life to teach me about trust.

Abbey was a German Shepherd cross who I literally found on the street, emaciated and very frightened. Now, this was in the days before I ever thought about becoming a dog trainer, so knew very little about canine behavior. A friend and I spent well over an hour trying to coax Abbey to come with us. Finally, we had an idea that worked. We employed my friend's German Shepherd to gain Abbey's trust and she hopped into the car with him.

Abbey was a broken dog; probably the result of abuse or neglect, or both. She didn't even know how to play. She was for the most part friendly and compliant, but she was fearful at the prospect of the slightest pain or discomfort.

I'm sure her physical health contributed to her demeanor, but something about Abbey was untouchable. She was deeply damaged emotionally. Her aggressive behavior was unexplainable and unpredictable.

As I said, this was long before I became a dog trainer, and I operated on old school philosophy that used compulsory methods instead of the reward-based methods I use today. I deeply regret that. Ah, if only I knew then what I know now ...

Abbey's lessons are those taught in hindsight. If given another chance, of course, I'd do things differently. I can't help but wonder if the compassion and understanding that I have now would have resulted in a better life for Abbey. This reminds me of a favorite quote by Maya Angelou: "Do the best you can. Then when you know better, do better." Because of my experience with Abbey, I can better help dogs by applying her teachings.

I recently had an animal communicator connect with Abbey. The communicator said that Abbey didn't 'speak' of the things I'd been feeling guilty about. She said she had had a happy life with me, and conveyed the words 'heart snuggles' to the animal communicator. What a relief for me. This is a great example of what animal communicators can do for the human side of the relationship. (More on this in later chapters.)

Finding Mr MoJo

Mr MoJo, who features on the front cover of this book, taught me to listen.

My best guess is that Mr MoJo was a Border Collie/Great Pyrenees mix, and a force of nature. He was highly intelligent; so intelligent that I believe he bordered on acquiring the power of speech. He spent the whole of our nearly 14 years together training me, and almost succeeded.

MoJo was an escape artist and a shape-shifter. That 70lb dog could fit through a hole in the fence the size of a Pomeranian. I once witnessed him use his teeth as wire cutters to open a hole in the fence, and then climb through. One morning, I awoke to a household minus one dog. In my search for the missing MoJo, I discovered the screen in the back door pushed out, with MoJo standing on the other side of the door. Now, here's the mystery. The window was open by about 4 inches only. Shape-shifting is the only explanation I could summon.

MoJo was a control freak. Thus, I called him the mayor of Aurora, the city where we lived. He conducted business and surveillance from his corner office in the yard. In addition to our city, he ran the household, announcing mealtimes, delivering wake-up calls, and broadcasting the arrival of the trash truck. His last moments on this planet are the most poignant moments of my life. He did speak very clearly that day.

Mr MoJo had been in a slow decline over the past several months. He was nearly 14 years old, but it was difficult to watch the once-vibrant force of nature slowing down, and his once dazzlingly bright light begin to dim. He was having mobility issues and his mental capacity faltered a bit. Wanting to give him the best quality of life in his last days, I called a veterinarian specializing in end-of-life cases for her opinion on his condition and an assessment of his quality of life. Her assessment was that MoJo's pain was under control and he was doing well under the circumstances, but his time was near. She said I could make the decision to help with transitioning within a few days, or she would support me if I made the decision to do it that day. I responded with, "Oh, no! Not today. I'm not ready!" I just wanted to know how to make him comfortable or what we could do to give him more time. "No! Not today!"

On further thought, I decided that the timing wasn't up to me, but ultimately up to MoJo. So I asked him, "What do you want to do, Big Guy?" The vet and I were sitting on the floor with MoJo lying between us. As I asked the question, MoJo's eyes brightened, reflecting perfectly clarity. He looked at me, then turned to the doctor and placed his paw in her hand. MoJo had spoken.

I gave the doctor the okay, spoke to MoJo for a few minutes, and then he got up completely under his own power and lay down exactly in the spot where his predecessor, Bob, had taken his last breath. Peacefully, MoJo closed his eyes and left us.

Finding Bob

Bob came into my life to teach me about humor and speaking out.

Bob was probably a Newfoundland mix, whose full name was Bob Barker. I gave him that moniker for the reason you might guess: he was quite vociferous. Bob was the absolute goofiest dog – yes, even goofier than Sadie. He always acted as if he was running the show. His boisterousness was akin to that of a politician who was always running for office. He would boss his brothers, Jude and MoJo, but only because they let him. He was, in their eyes, full of hot air.

Finding Jude

Jude came into my life to teach me about drama.

Jude was a purebred German Shepherd. A friend and colleague informed me that he'd been relinquished to the veterinary clinic where she taught dog training classes. He was, in my opinion, one of the most handsome German Shepherd pups I'd ever seen, and he was already enormous at just 6 months old. Oh, but he was a classic drama king, who could have had a starring role in the most theatrical daytime TV drama. And, what a whiner! That's how he got the name, Jude. Being a lifelong Beatles fan, I often name my dogs after a Beatles' song title, and, because of Jude's excessive whimpering, I always said that we had to take his sad song and make it better. Ultimately, Jude trained me to play a supporting role in his lifelong melodrama. I often willingly catered to his drama and martyrdom: I didn't mind, really …

Finding Penny

Penny came into my life to teach me about sensitivity.

Only four days after my German Shepherd, Jude, left this world, I received a text and photo of a stunningly beautiful dog. My friend's text said that she knew of a German Shepherd up for adoption, and asked if I was interested. Being such a short time since Jude's passing, I immediately rejected the idea. I needed time to grieve, so I thanked my friend and closed the file, telling myself that if this dog was meant to come to me, she would.

Several months later, it was time. I contacted the person with the German Shepherd and found that she still had the dog. I asked about her story, and the woman told me that she had taken in Penny after she had experienced a traumatic incident, which manifested as severe fear aggression. After training and progress in rehabilitation for her aggressive behavior, Penny was ready to find a permanent home. Understandably, the woman was selective about whom she'd allow to adopt Penny, but, with me, she felt she'd found the perfect match for her. As a dog trainer, I could finish and maintain Penny's rehabilitation. And so it was.

Penny Lane came into my life to teach me about sensitivity, tuning in to energy, and overcoming fear. Her trauma taught her these abilities, and now she's passing on those lessons to me. Penny has shown me the patience required to conquer fear. She taught me the importance of allowing her to do it on her own terms, and is a role model to many of the fearful and reactive dogs I work with. She has even assisted me with their training. Moreover, she is a role model to me.

Penny is now retired from working with reactive dogs, and her current mission is to teach me to read energy and to cultivate a deeper communication with animals. She is by far the most sensitive dog I've ever known. She recognizes when I'm frustrated, angry or afraid, even before I do, and teaches me that being connected to the energy of the Universe means being connected to myself.

Finding Jett

Jett, a Great Pyrenees mix, came into my life to teach me about trusting my intuition and taking life easy. His predecessor, Mr MoJo, sent him to me, I'm sure of it.

Mr MoJo was indomitable. I feel that since he took charge of my life during his time here on Earth, he continued to do so after his departure, and therefore he appointed his successor.

It was time for another dog. I've always been attracted to big, fluffy, goofy dogs, but at that point in my life, I thought I should downsize. I pondered what mid-size breed would be best. For weeks I monitored rescue sites, but no one was 'speaking' to me. Then, I saw him. Someone fostering for a rescue group I've worked with posted a photo of Jett. He was a big, fluffy, goofy dog. Nonetheless, the voice in my head said, 'That's my dog!' I read his profile and immediately filled out an application.

I already knew that Jett was the one, but this was validated when I emailed his photo to my highly intuitive friend. He said that when he opened the photo of Jett, he'd had the Elton John song *Benny and the Jets* stuck in his head. I replied, "No, it would be Penny and the Jett!" to which he responded, "This is SO your dog!" which I took as further confirmation. Before I had even met him, Jett began teaching me to trust my intuition.

Jett was also sent into my life to teach me to relax. Immediately after moving in, he took over my bed. He shares it nicely, but there's nothing Jett likes more than napping there. He can be jetting around the yard one minute and stretched out in slumber the next – a lesson in letting go that I can learn.

Further teaching me how to relax, Jett shows me how he dislikes being under pressure and his aversion to chaos. I can be brutal about pressuring myself, and I have a strong dislike for crowds. Jett reflects lessons I need to learn and I trust his teachings.

All of my dogs taught me to follow my heart, as they always followed theirs. I followed my intuition with Penny and The Jett, and they show me how trusting in myself leads me on my true path.

THE STORY OF ROREY

We all have tales of how we found our dogs ... how we found each other. This is the particularly poignant story of how my friend Courtney's dog, Rorey, found her.

Letting in the dog

Courtney had been working for the Humane Society in California, which had taken in a litter of one-day-old pups. The mother dog was too ill and emaciated to feed her babies, so the staff bottle-fed them. Courtney was particularly drawn to one of the three puppies, who she named Grizzley. Courtney nursed and cared for this infant pup daily, even carrying her around in her shirt, cuddling her to keep her warm. They formed a strong bond, and, knowing that it was meant to be, Courtney adopted Griz.

At 5½ months old, Griz was scheduled for routine surgery, but blood work and testing showed that, sadly, Griz had major problems. Among other issues there was a hole in her liver, and she had only one functioning kidney. Doctors gave her just 2 weeks to live.

Courtney promptly made a bucket list for Griz and carried it out to the letter. She took time off work and devoted herself to Griz's last days. Daily, she took Griz to her favorite beach for joyous romps, and when Griz was too weak to run on the beach, Courtney carried her there, and Griz was happy to just sit gazing at the surf and the horizon. Then Griz told Courtney that she was ready to go and transitioned to spirit.

After receiving Griz's ashes, Courtney gathered a group of friends to help scatter them on the beach that Griz so loved; the very beach where she'd had her final romp. When the observance was complete, Courtney felt as if she'd scattered the pieces of her broken heart along with the Griz's ashes. As tears were still streaming down her face, Courtney looked up to see a 12-week-old puppy racing full speed toward her. No one was with the pup; she had simply appeared out of nowhere! The puppy picked out Courtney from the group, made a beeline toward her, and enthusiastically climbed all over her, licking away her tears.

Courtney saw that the puppy was dragging a leash on which was printed the name of a rescue group and the words, 'Adopt me!' Courtney looked up the group's contact information online, called it, and discovered that the pup had escaped from an adoption event being held nearby. Now, when I say nearby, I mean a drivable distance, not a reasonable distance for a 12-week-old puppy to run. Rorey had travelled half a mile to find Courtney!

You know the ending to this amazing story, of course: Courtney adopted Rorey. She firmly believes that Griz sent her directly ... and it's hard to argue with that.

Rorey is a natural at sensing emotion; I've seen it for myself. I was telling Courtney an emotional story about one of my dogs when she said, "Look, Rorey has taken her attention from me and is looking at you." Then Rorey came and sat between my feet, leaning into my leg. She was comforting me.

There is no question about why Rorey came into Courtney's life. This amazing creature is currently in training as Courtney's service dog. She knows her purpose and will carry it out perfectly.

I believe that energies attract similar energies. I believe, therefore, that we attract everything and everyone in our lives. We've attracted every person, animal, job, possession, relationship, and experience. All show us a mirror. Our love for our dogs has attracted them to us to demonstrate how the Universe works, and how its power is within us.

Canine communication: it's both intellectual and intuitive

IT IS QUITE POSSIBLE THAT AN ANIMAL HAS SPOKEN TO ME AND THAT I DIDN'T CATCH THE REMARK BECAUSE I WASN'T PAYING ATTENTION
E B White, Charlotte's Web

I always like to say, 'Depend more on your relationship with your dog than you do on your equipment,' which means engage with dogs in the training process rather than depend on leashes and collars (especially corrective collars) or wish for a magic pill to solve a problem.

There are some situations that training can't resolve, and in these cases, we should not automatically blame the dog and place unrealistic expectations on him, but need to get creative with our solutions. I once met a man who wanted to leave his mail-eating dog muzzled all day. I suggested that he install a mailbox on the front porch instead of having the letter carrier drop the mail through the mail slot. The problem was solved without putting the dog in an uncomfortable and unsafe situation.

We get a lot further in training when we open the lines of communication and open our hearts, our minds, our ears, and our eyes. It's equally important to listen to what the dog is saying with his actions. Why is he pulling on the leash? It reminds me of the Tom Petty and The Heartbreakers with Stevie Nicks' song, *Stop Draggin' My Heart Around*. We get frustrated with the pulling and it feels like our emotions are being dragged around. Well, maybe so, but only because we're allowing it to happen. The dog is merely pulling because he's excited to get to his destination, not to cause us frustration.

'Successful' walks take cooperation. I imagine that your dog does not think that dragging you after him makes for a successful walk. He's frustrated, too. Maybe we should reframe our thinking: let's not walk the dog; let's walk *with* the dog. Thus, we can have conversation and cooperation instead of a war of wills.

As humans, we are very good at talking to dogs, or maybe it's more accurate to say that we talk *at* dogs, and I have written this book to enlighten readers about both sides of the conversation. This chapter will focus on how we can listen more than we talk. I like to point out that the words 'listen' and 'silent' are composed of the same letters. If we are silent, we are better able to listen.

A big difference between dogs and humans is that most (behaviorally healthy) dogs can spontaneously move on from adversity to be delighted in something as simple as the magic of a leaf blowing by. We might learn from their ability to be present.

LISTEN TO YOUR EYES; HEAR WHAT YOU SEE

Understanding canine communication goes beyond the workings of academic knowledge and science, and to effectively communicate, we must appreciate both the intellectual and intuitive components of this two-way conversation. To let in the dog, so to speak, it's essential to embrace how dogs think and learn.

Let's begin the dialogue with a little scholarly understanding of canine communication, and from there, we'll look at the merging of hearts and minds into understanding. This is where we begin to see the whole picture and hear the entire story.

THERE'S AN APP FOR THAT

New social media platforms are emerging daily, it seems, with ostensibly new, easier ways to communicate. So much so, we could be at risk of losing facetime and the ability to pay real attention to the conversation. We spend hours scrolling and scanning our news feeds, and here we are communicating on a rather shallow level where much can be lost in translation. So much can be misinterpreted and misunderstood without the benefit of voice inflection and body language.

Dogs don't have electronic social media, but they are social animals who use scent, vocalization and body language to communicate. To enable a peaceful, successful co-existence with dogs, we need a deeper understanding of canine social interaction. It behooves us to accurately interpret what dogs are saying to each other – and to us.

Instagram became popular largely because people appreciate visual communication. A picture is worth a thousand words, right? While dogs relate using vocalization and scent also, body language is their predominant form of communication. We humans are largely left out of the scent part, so let's concentrate on understanding the vocal and the visual.

Social animals like humans and dogs have highly evolved ways of communicating. A wide array of varying components comprise canine body language: dogs use facial expressions, tail carriage, vocalization, and overall demeanor to communicate with each other, and with us. It is helpful to pull apart these signals into individual components in order to understand the often subtle language behind them. We will have a look at the individual signals, but remember, when reading dogs we must look at the whole picture (not just the wagging tail) because dogs use these signs in concert and in context.

Dogs primarily use communication to signal intent. While scent is one of a dog's primary forms of communication, by and large they communicate through body language, with vocalization being a lesser part of their communication. These signals may appear random to us but they are not: they serve to relay a dog's internal state or are a purposeful attempt to tell us something. Many of these signals are used to negotiate disputes, navigate potentially conflictive situations, and avoid conflict altogether.

Canine body language is a window into the minds and feelings of dogs. In understanding the language, we will build a stronger relationship because our dogs will feel understood, and understanding builds trust.

Study canine body language (see the recommended reading list.) Learn to recognize the commonly-known signals (tongue flick, look away, paw lift) as well as lesser-known and more subtle ones (sniffing the ground, blinking.) Discover the meaning of displaced behaviors, such as yawning, sneezing, and scratching. These are normal, familiar behaviors done out of context when a dog needs comfort or to escape. Learn about ambiguous behaviors where the dog's actions do not necessarily mirror his intentions. Here we must rely on the whole picture of the dog and the context wherein the behavior occurs.

There is no one as deaf as he who will not listen (Yiddish proverb)
Before we delve further into canine body language, let's discuss the notion of dominance. Dominance in the context of how dogs establish rank and understand hierarchies can be a flashpoint topic in the world of dog training. In order to stay on-point, let me give a straightforward explanation: dominance simply means to exercise the most control in a social interaction. It is better described as a state, not a trait, and it is contextual.

Dogs do form hierarchies and establish rank within these. However, dominance is not recurrently established or enforced through physical conflict and aggression. If you observe canine social interactions, you'll see voluntary deference employed in various situations. In appropriate interactions, role reversals take place, with dogs exhibiting deferential behavior alternately.

Canine social hierarchies tend to be fluid. Communication involves the use of body language such as the dominant postures (assertive gestures) and submissive signals that are described below.

In the following descriptions, the term 'dominant dog' refers to the higher ranking dog, or the dog exercising the most control in a particular interaction.

Now we're ready to take a closer look at dog body language by breaking it down into individual signals.

TONGUE FLICK

This signal can be very subtle and quick. Sometimes the tongue is barely visible outside the mouth, and other times it is extended far enough to lick the nose.

Dogs may lick their lips after eating, but in the absence of food, dogs use licking the lips to signal intention. He could be saying that he means no harm as he approaches. It may be used to tell us or another dog that he is feeling nervous and wants us to calm down. When it is a displaced behavior (ie in the absence of food), a tongue flick indicates that the dog is experiencing anxiety. The behavior may be seen in dogs who are nervous when visiting the vet's office or the groomer.

SNIFFING THE GROUND

Sniffing the ground can be a quick, sweeping, downward motion or a prolonged activity that continues until a situation is resolved. Dogs use this signal often when in groups. Dogs are programmed to use their noses. It's an enjoyable activity, so sniffing often means they are merely gathering information from the environment. Additionally, dogs use sniffing the ground as a calming or intention signal in order to calm other dogs or humans. The cause for sniffing depends on the situation. For instance, if a dog abruptly stops playing with another dog, suddenly ignores him and sniffs the ground, he may be telling the other dog to calm down.

TURNING AWAY OR TURNING THE HEAD

Turning away is a universal signal among dogs, and they will turn their heads or whole body away from a perceived threat. The intensity of this signal can range from holding the head to the side for a long time, a quick turn of the head and back, or simply averting the eyes. Some examples of when dogs use the turning away signal are: other dogs or people approaching too quickly or head-on; someone seems angry, or the dog is taken by surprise. The use of this signal is an effective way for dogs to avert conflict.

PLAY BOW

This signal is used as an invitation to another dog to play, and as an intention signal. If the play bow is bouncy with the dog moving his front feet from side to side, it is most likely an invitation to play.

Sometimes the play bow is intended to calm another dog. A dog who is fearful of another dog may demonstrate this, as the play bow would serve to diffuse a tense situation. A play bow can be an active display of inefficient motion or it can be still and prolonged. If the dog is standing still, the play bow may be intended to calm another dog.

WALKING SLOWLY OR 'FREEZING'

Chaos and speed can cause anxiety with many dogs. A dog who feels insecure may move slowly and deliberately, often freezing when approached too quickly or if feeling threatened. Often, dogs will use these two signals together: they might walk slowly, freeze, and then walk slowly again.

WALKING IN A CURVE

In polite, friendly greetings, dogs naturally do not approach each other head-on.

Instead, the body is loose and they walk in a curve toward each other, avoiding direct eye contact. They then proceed to sniff each other's behind.

We never want to force dogs to meet each other head-on, especially when they are on-leash. This can cause great anxiety, which could lead to a confrontation. Allow the dogs to curve toward each other at their own discretion. The more anxious a dog is, the wider circle he should be given.

BUTT SNIFF
A butt sniff is what we call a doggie handshake. Sniffing is how dogs get information about each other. Done politely, it does not invade the other dog's personal space by keeping the nose at a respectful distance and the sniffing is kept brief.

PASS BY
A polite pass by is not a head-on encounter. The dog does not block the oncoming dog or display an aggressive posture (leaning forward, stiff, and/or on tiptoe.) A civil pass by does not invade the space of the oncoming dog, as it gives him a wide enough berth. Appeasement signals are given, such as paw lifts, and the dogs show a neutral, relaxed posture. Dogs passing by are alert, yet calm.

Humans on the other end of the leash can compromise pass bys. If we become nervous, we communicate our anxiety directly through the leash. A tight leash can also cause frustration for the dog. If we force our dog to strain on the leash, he presents a more aggressive posture to the approaching dog (leaning forward with stiff body). Keep leashes loose for successful pass bys.

SITTING OR LYING DOWN
Dogs may try to calm humans or other dogs by sitting when approached. An even stronger signal is a dog sitting with his back to us when we advance too quickly or sound angry. Additionally, a dog may lie down and turn his head away from the perceived threat. Humans tend to interpret it to mean the dog is ignoring us when actually he is trying to calm us.

SPLITTING
When a dog feels tension between other dogs or dogs and humans, he may physically place himself between them, separating them and helping to prevent escalating tension or conflict. We may see dogs use this strategy when two people hug or when children roughhouse. Herding dogs tend to use this signal a great deal, but it is seen in other breeds as well.

CHIN OVER
Dogs will often place their chin over another dog's shoulders or back as a signal of intent. It is frequently a status-seeking behavior to determine who is the higher ranking dog, or can be used as a threat. However, a chin over can also be used as an invitation to play. To determine a dog's intent, consider the context of the situation and reaction of the other dog to the signal.

PAW LIFT
A paw lift can indicate curiosity, uncertainty, insecurity, submission, stalking, or anticipation (ie waiting for the ball to be thrown). The paw lift is a signal often

used in conjunction with freezing when intended to show caution. Frequently, dogs use this signal when meeting for the first time to indicate that they pose no threat to the unfamiliar dog. Dogs may use their paws as a signal of appeasement or deference in greeting.

Paw over
As with a chin over, dogs will often put their paw over another dog's back or shoulders as a signal of intent. This signal can be a status-seeking behavior to determine rank, or it is sometimes seen as an invitation to play.

Facial expressions
A dog's facial expressions include the ears, eyes, eyebrows, and mouth. Wolf and dog facial expressions are similar, but the differences in features among breeds can make a dog's facial expressions more difficult to read. For instance, brachycephalic dogs (those having a relatively broad, short skull and short muzzle) such as English Bulldogs, have faces that are harder to read because the corners of the mouth are more drawn back in a neutral, resting face than that of, say, a German Shepherd. Furthermore, the wrinkled face of a Pug or an English Bulldog makes for an invariably furrowed brow. Reading facial expressions is not as well defined as signals such as play bows and paw lifts, leaving more to interpretation.

Eyes and eyebrows
Eyebrow expression underlines what the dog's eyes are saying. A dog displaying a dominant posture will have a well-defined brow; a furrowed or wrinkled brow can be a sign of stress. The brow is much more difficult to read in breeds such as the Great Dane and Shar Pei because their faces are naturally wrinkled.

Eyes
Dogs' eyes vary in shape and size, with some being very round and others more almond-shaped. Within the limits of these shapes, all dogs can widen or narrow their eyes, and change the intensity of their gaze. Eyes that are wide, showing the whites (often called whale eye), and dilated pupils indicate that the dog is stressed.

The muscles around a calm, happy dog's eyes are relaxed and the face has an open, full appearance. A hard, fixated stare signals a threat, which is most often directed at other dogs. The stare is very focused and intense, and the body is stiff. However, many dogs learn that looking directly at people can be okay; even pleasant. This is done with a relaxed, friendly expression.

Blinking
Blinking the eyes can be an appeasement behavior used to show submission. Dogs will use blinking with humans if feeling stressed or confused, or when being scolded.

The ears
The vast diversity of ear types among dog breeds makes reading them a challenge. Size and shape dictate their effectiveness in communication. No matter what the ear type, a calm dog holds his ears relaxed and natural. On alert, the dog will hold the ears higher, and direct them toward the point of interest.

Friendliness displays a relaxed, slightly laid back ear carriage, whilst flattened ears are a sign of fear or submission.

THE MOUTH

A dog's mouth speaks volumes, but with more than just sound. The position of the lips and the commissure (the corners of the mouth) are important cues to look out for. A dog at ease will often have his mouth closed and relaxed, or slightly open, tongue lolling, and panting. The lower jaw is relaxed and there are no ridges around the lips.

A short commissure, which exposes the teeth and displays a wrinkled nose, is an agonistic pucker, which is a warning signal. It is often accompanied by a growl, furrowed brow, and flattened ears. However, the agonistic pucker can have other meanings, depending on context. It may also be used as a defensive, territorial, or submissive/aggressive display; when a dog is feeling uncertain, be a warning to another dog to go away or calm down, or when guarding a resource.

A display of submission usually involves a closed mouth with lips slightly pulled back. There may be tongue flicks. Occasionally, dogs display a submissive grin, which, as the name indicates, shows submission. The lips are vertically retracted, showing the front teeth – the canines and incisors. Laid back ears, squinty eyes, and a generally submissive body posture often accompany the grin. Submissive grins are easily mistaken for aggression because the teeth are bared.

A dog with intent to aggress will also retract the lips vertically to expose the teeth. The nose will be wrinkled. Commissure showing the lips drawn back horizontally showing both the front and back teeth is often indicative of fear. A dog about to bite will open the mouth pulling the lips up and back with the teeth exposed.

VISUAL SIGNALS

The tongue

The tongue of a dog who is panting because he is under duress is held tighter and flatter at the end (and even curled upward) than a dog who is panting because he's happy or just hot. The stressed tongue is often called a spatulate tongue and is held in place; not simply left to fall out of the mouth as with a relaxed dog.

Licking

Mother dogs communicate with their puppies by licking them, which stimulates the pups to start breathing: it's also how she cleans them. In the wild, puppies will lick adult dogs' mouths to encourage them to regurgitate food after the hunt. Licking in adult dogs is most often used to show friendly submission. They use it with us and with other dogs most often as a sign of affection.

Many dogs lick themselves (often their paws or legs) to relieve stress, and sometimes they will lick objects or the floor. Licking can have a calming effect because it releases endorphins; thus relieving stress. Chronic licking can mean the dog is bored or anxious. It can also be indicative of skin problems such as allergies, or a signal that the dog is in pain.

Drooling

Profuse drooling in the absence of food is an indication that a dog is stressed, even though he may not appear to be stressed, showing few other signs. If he is drooling excessively, it's likely he is feeling anxious or nauseous. (Note that some breeds, such as Basset Hounds, are natural droolers, so here, drooling is not necessarily a sign of stress.)

Tails

Reading tails is difficult, in that the natural carriage differs greatly between breeds. Greyhounds are one of a few breeds whose tail set is naturally down, tucked slightly between the hind legs. The saber tail of a German Shepherd is much easier to read than the corkscrew tail of a Pug. The sickle tail of a Husky or the flagpole tail of a Beagle is easier to interpret than the docked tail of a Rottweiler or the bobtail of a Welsh Corgi.

The idea that a wagging tail is always that of a happy, friendly dog is a dangerous conviction. Dogs wag their tails for a multitude of reasons, including happiness and alertness. Surprisingly for some, a wagging tail can be an indication of an aggressive dog: his tail can be wagging all the while he is barking and lunging at a perceived threat. A dog feeling affable and relaxed will hold his tail in a natural position. A happy dog will wag the tail gently; a very happy dog will wag exuberantly, even doing a whole-body wag with lots of curves in the body's motion.

Nervous dogs will tuck their tail between the legs, but can include a wag, although this is often more rapid than when the dog is relaxed. If the dog is extremely nervous, the tail may be tucked so high as to touch her belly. A rapid vibration at the tip of the tail held perpendicularly is a sign that the dog is anxious and stressed. A dog on alert will hold the tail high, rigid, and still. A tail held high and stiff with slow, intentional, back-and-forth movement signals a threat. The whole body is tense, the weight is on the front legs, and the feet are positioned squarely, ready for fight or flight. Slow, large amplitude, catlike tail swishing is also a warning sign.

Hair

Increased sudden shedding, and/or excessive dander are reflexive stress signals, though not intentional or consciously controlled. The hair is raised at the withers (the base of the neck above the shoulders), or sometimes raised in a ridge all along the spine. Some dogs raise the hair only at the base of the neck and the root of the tail. Raising the hair is called 'piloerection' or more commonly termed 'raising the hackles.' Often, we interpret this action as an aggressive gesture, which it can be. However, dogs may raise their hackles any time they are in a state of arousal. Raised hackles can mean that a dog is fearful, anxious, angry, unsure, surprised, or very excited about something; the human equivalent is goosebumps.

Stretching

Dogs often stretch as a greeting; usually only with people and dogs they like and are comfortable with. The exercise is similar to a play bow, but more relaxed and protracted. It is not an invitation to play followed by pouncing or incitement to be chased. Some dogs will do a forward stretch, where they stretch out their back legs behind them. Some dogs will do the two stretches in sequence: first the front stretch (like the play bow), then the rear stretch (legs stretched out behind).

Posture

Having looked at individual body parts to tell what a dog is feeling and saying, let's have a look at the bigger picture, which is the carriage of the entire body where individual signals come together as a whole.

Neutral

A neutral body posture shows relaxed muscles with the weight evenly distributed on all four feet. Head and tail carriage are normal and relaxed.

Anxious

A frightened dog will be hunched in an effort to look smaller. The head will likely be lowered, and he will possibly cower on the ground. The tail may be tucked under slightly, or held up tight to the belly.

Cautious

A cautious dog may hold his body low, approaching tentatively. His weight may be on his back legs in order to make a hasty retreat if deemed necessary.

Submissive

A submissive dog will lower his body or cower on the ground. He will display a cautious body posture on approach.

Alert or aroused

A dog on alert holds his head and tail high to appear larger. His muscles are tensed, ready for action, and he's standing squarely on all four feet; usually with the weight centered more on the front feet, ready to move forward.

Aggressive

The aggressive dog looks similar to the alert or aroused dog, but will be displaying other signs of aggressive intent, such as bared teeth, wrinkled nose, and growling.

VOCAL COMMUNICATION

Barking

Dogs bark as an alert to perceived dangers and to warn of intruders. Dogs also bark to get our attention and when they're bored. During conflict, a more varied tone of bark indicates submissiveness. A bark that invites play will be higher pitched and repetitive.

Howling

Dogs instinctively howl as their wolf-like ancestors did. Wolves howl to give information about territorial boundaries, and to bond with pack members. Howling is the wolf's version of GPS, used to locate and communicate with individuals. Howling is more effective for long distance communication because it is sustained and protracted, and can travel for several miles.

Dogs might howl when they're lonely. Howling is contagious among dogs. If they are howling to communicate, then the more howlers, the more volume; therefore the sound carries farther.

Whining

Whining is typically associated with submission, fear, insecurity, and loneliness. It can also convey the distress of pain and anguish, and/or anxiety due to stress. Yet, some dogs will whine when they are excited, such as during greetings. Whining may be used as an appeasement behavior when interacting with people and other dogs. In this context, other submissive body language may be displayed, such as cowering, tucked tail, lowered head, and averted eyes. Some dogs whine for attention from their owners or to solicit rewards, food, or other desired objects.

Snarling

Snarling is a signal of aggression, in which the lips are drawn back to expose the teeth, and often accompanied by a deep, guttural growl. A snarl with a forward commissure is an offensive or dominant signal. A snarl with a retracted commissure is a defensive or submissive signal.

OTHER BEHAVIORAL COMMUNICATION

Marking

Dogs use urination and defecation to mark territory or objects by leaving their scent on it. In the wild, this is necessary because of competition for resources and mates. Marking also says to other dogs, 'I passed this way.' Many dogs mark visually by scratching the ground, often after urinating or defecating.

Body contact

A gentle bump with the muzzle to another dog or to a human hand or leg is a friendly gesture showing acceptance. It is an appeasement behavior, which also signifies submission. A soft push with the hip also indicates friendliness, and is often used in greetings, demonstrating trust and friendliness where the dog intends no harm. Lying together with bodies touching can strengthen the bond between individual dogs and help them feel secure. A dog may snuggle up to you for security or for the company, but done more forcefully, it can be a status-seeking space-invading move.

Muzzle grab; muzzle to throat

A puppy's early education includes being grasped around the muzzle or head by his mother, from which the pup learns submission. It can also be seen in play and social interactions between adult dogs. Muzzle grabs can be a display of dominance, with the higher status dog doing it to the submissive or lower-ranking dog. A nudge to the throat with the muzzle is a pacifying behavior of a submissive dog to a higher-ranking dog.

Pawing

Pawing is where one dog touches or strikes another dog, a person, or something with his paw. A dog may also paw the air. This is a gesture made with the intent to pacify, as in a paw lift. It's often used when a dog is confused or frustrated.

Shake off

A shake off communicates non-threatening intent, and means the dog is literally shaking off tension, energy, adrenaline. A shake off moves the dog from a reflexive state to a cognitive one, which says to us and to other dogs that he is thinking, communicating, and is more predictable than he was in the reflexive state.

Displacement behavior

Displaced behaviors are those in which a dog engages when he needs comfort or to escape, and are indicative of a dog feeling conflicted or stressed. They are normal, familiar behaviors that are done out of context, which help achieve a sense of security in uncomfortable situations. These behaviors are used when a dog is suppressing what he really wants to do. Displacement behaviors are contextual, as dogs do them for other reasons. For example, a dog will scratch when he has an itch. However, if he suddenly begins scratching during a training session, it might be a sign that he's confused, frustrated, or feeling pressured.

Yawning

Dogs yawn when they are tired, of course. Additionally, it's an appeasement behavior often used by a dominant dog to show friendliness to a submissive dog, and vice versa. Yawning can be a displaced behavior done when the dog is feeling nervous, confused, or frustrated.

Scratching

Dogs might use scratching as a displaced behavior to calm themselves when feeling confused or frightened. They may also use it as an appeasement signal when feeling threatened.

AMBIGUOUS BEHAVIOR

As mentioned, dog body language, actions, and vocalizations can have an assortment of meanings, and a dog's actions do not necessarily mirror his intentions. Therefore, we must rely on the whole picture of the dog and the context wherein the behavior occurs.

For example, a dog who growls at another, with the vocalization followed by a play bow, a relaxed mouth with lolling tongue, and a paw lift, is inviting the other dog to play. A growl during a game of tug is part of the game. Conversely, a growl with a stiff, forward posture, and a hard stare is a warning, which, if unheeded, could escalate.

To further muddy the waters, dogs can feel conflicted and give mixed messages, so we must then make our best guess as to what the dog is feeling. Simply knowing that he is feeling conflicted, however, is a big step toward helping him through the uncomfortable situation.

Learning to communicate with dogs builds strong, trusting relationships, and engenders patience, relieves frustration, and develops confidence at both ends of the leash. I dedicated a whole chapter to dog body language in my book, *Taking the Lead without Jerking the Leash, The Art of Mindful Dog Training*. Two other excellent resources for understanding communication are *Canine Body Language, A Photographic Guide* by Brenda Aloff, and *On Talking Terms with Dogs, Calming Signals* by Turid Rugaas.

THE QUIETER YOU BECOME, THE MORE YOU CAN HEAR (RAM DASS)

I am fascinated by the remarkable similarities in the way humans and canines express themselves. We have similar facial expressions, such as a furrowed brow. We can show disinterest by walking away or diverting our eyes. We express surprise with wide eyes, and indicate curiosity by tilting our heads. Why, then, do we sometimes have so much trouble communicating?

To break down barriers, we must let go of what we believe to be our differences. The biggest barrier is the human ego, and we must let out the ego in order to let in the dog.

Humans have such cluttered minds, and in Buddhist principles this is called 'monkey mind,' meaning an unsettled, restless, or confused state. In both dog training and animal communication, we must ensure purity in the transmission of information and thought. We need to be a clear conduit. The busy, cluttered mind clogs lines of communication, and casts a shadow upon the conversation. The purest conversation is held when we are being mindful instead of having a mind that's full.

Most simply stated, communication is a message sent by a sender and understood by a receiver. Humans predominantly use conscious visual and audial communication, whilst dogs obtain and share information about the world predominantly through scent. Humans largely overlook other forms of communication such as scent, electronic impulses, and chemicals. Elephants communicate in part through low frequency vibration via the ground that is picked up through their feet. Killer whales, unlike any other mammal, create and receive highly specialized sounds through the contours of their skulls. Bats, whales, and dolphins can communicate by sonar.

We consider these to be 'lesser' forms of communication because they are not human, but an animal's behavior will tell us what he or she needs; it would be demeaning to ignore the conversation.

Words are all too often an inadequate tool for measuring thoughts. We can hear what dogs are saying by carefully observing them. Dogs wear their hearts on their sleeves, and we can see what emotions they are feeling. As we know, dogs share many of the same emotions as humans. Shared emotions magnetize us and bond us.

THINKING IN PICTURES

Now that we have some basic facts about canine communication, let's delve a little deeper into the conversation.

In Temple Grandin's book, *Thinking in Pictures: And Other Reports from my Life with Autism*, Grandin describes how she connects with animals because she understands that they see in pictures. She states that anything creating visual contrast will attract an animal's attention. For instance, reflections can be frightening to an animal; darkness or low light can cause an animal to be fearful or hesitant – they want to see where they are going just as we do. The opening words to chapter 1 of *Thinking in Pictures* are, 'I think in pictures. Words are a second language to me.' Grandin credits her ability to visualize with helping her to understand the animals she works with.

MEET ME WHERE I AM

Communication is the bedrock of successful relationships. In training, I say that we must meet our dogs where they are: where else is there, after all? We miss the meeting place when we let words, expectations, egos, and over-inflated goals get in the way. Effective communication with dogs involves being mindful, being present, and having a full understanding of how dogs think and learn. In addition to that general understanding, we must know our own dogs and have realistic expectations of them. We have a duty to understand and communicate with the dog we have, not the dog we wish we had.

FINDING EACH OTHER

The Universe does not necessarily give us the dog and experiences that we *want*, but it does give us the dogs and experiences that we *need*. In Chapter 1, I described how I found each of my dogs ... or maybe I should say how they found me? Either way, we found each other. Every dog who has come into my life has done so with a purpose, whether or not I realized it at the time. Every one was and is a teacher.

BEING REALISTIC ABOUT WHAT A DOG CAN DO AND CANNOT DO IS AN ACT OF LOVE
Suzanne Clothier

GREAT EXPECTATIONS

A great disparity exists between our expectations of dogs and their canine reality. Expectations can lead to disappointment and resentment, but if we open our minds and the lines of communication – if we're willing to be flexible and accepting – we can find equal ground on the path to a happy co-existence. It's helpful to pause for a moment on our journey to adjust or ease our expectations, and to define individual success. Instead of relying strictly on management or corrections when problems arise, we can get out in front of the situation and see around the corner in order to anticipate and plan. We can expect to be patient and flexible, and we can expect the unexpected.

EXPECTATION AND EXCEPTION

We expect to find patterns and absolute answers even where none exist. We expect dogs to seek dominance, except when they don't. We expect dogs to come when called, except when they don't. We expect dogs to pay attention, except when they don't. The words expect and except contain the same letters. Let this remind us to *expect* the *except* part of life with dogs.

There is a fine line between hope and expectation. Expectation is looking toward an outcome with attachment and judgment, and could even be termed premeditated resentment. Hope is looking toward an outcome without attachment. Hope is compassionate. If we recognize the difference between hope and expectation, we can detach from the outcome. We can stop judging success by how the results fit our expectations. We can expect success no matter what it looks like. There is hope. Communication, understanding, and realistic expectations allow both dogs and humans to participate in the relationship. As such, we do not suppress the uncontainable spirit of dogs, but join them.

LISTEN AND LEAD

The greatest leader is the one who says the least and listens the most. This is true of leaders of dogs and humans. Think of authority figures who bellow commands constantly in order to gain control: ultimately, their subordinates learn to ignore and avoid them. Now, think of a softly-spoken authority figure. If on the rare occasion he or she raises their voice, they are more likely to receive compliance and respect – without fear.

The best leaders don't actually lead anyway; they guide. Furthermore, followers of benevolent, compassionate leaders retain a sense of control in the situation. They feel valued because there is a sense of participation as opposed to domination. Trust is engendered.

It's easier for humans to bark orders than it is to listen and observe. Sometimes, I invite clients to cue their dogs for a behavior without speaking. Oftentimes, I ask clients to have their dog walk beside them without using the leash. I'm told that this is the hardest thing I've ever asked of them.

STORYTELLING: TO GIVE THEM A VOICE, GIVE THEM AN EAR

With every potential client inquiry, I get to hear a story. People love to talk about their dogs. However, I must listen with an open mind and heart and remember that it is a story. It's a narrative (and an opinion) told by the person, not the dog. Frequently trainers hear, 'I'm sure that he was abused before we adopted him. That's why he acts that way.' This may be true, but only the dog knows the real story. There are varied reasons for a behavior, and the bottom line is that I want to observe and listen, with both heart and mind, to what the dog is saying. He's not going to tell me his life story, but he will tell me what he wants or needs in the moment.

I once told a client with a reactive terrier that her dog just wanted to feel safe on his walks. I could see the relief in her face: her story and her opinion about her dog's behavior immediately dropped away, and she thanked me for telling her about her dog's needs. Sometimes, we have to get past our own storytelling so that we can hear our dog's tale.

SCIENCE MEETS PSYCHIC

Letting in the dog is about finding the space that embraces academic knowledge and instinctive communication. It's about using heartspeak and mindspeak instead of mindless chatter and reflexive responses.

I define mindspeak as the language of the intellect. Heartspeak is the language of the heart. In the following chapters, we will explore the intuitive side of communicating with canines: we'll explore how to be with dogs in that space between scientific methodology and the psychic realm ... that place between intellect and the intuitive.

ANIMALS ARE LIKE PEOPLE BECAUSE PEOPLE ARE ANIMALS
Barbara T Gates, Kindred Nature

Are you thinking what I'm thinking? Am I thinking what you're thinking? Do you know what I'm thinking? Do I know what you're thinking? I think I'm confused!

We are often cautioned about attributing human characteristics or behaviors to dogs (anthropomorphizing), but this isn't always so far off the mark. If we understand canine communication, we can properly interpret what dogs are saying and know when anthropomorphizing is okay.

There's a meme that goes around social media showing a Beagle puppy hanging his head, cowering, and sad-eyed. It says, 'This is what guilty looks like.' Actually, this is what fear looks like. What we interpret as guilt is actually the dog's fearful reaction to the human's displeasure.

We don't really know if dogs feel guilt, but obviously humans are bad at reading it. We do know that dogs feel fear, and the pup in the meme is showing his fear of being punished. When we interpret canine behavior assuming that dogs understand the human construct of right and wrong, we have a problem. To feel guilty, a dog would need to understand right from wrong, as we do. We can teach them rules and boundaries, but to teach them that getting into the trash is 'wrong' is a bit much to expect. Dogs know how to be dogs. They are opportunists who know how to survive. If left alone with tempting trash, they aren't going to stop and think that getting into it would be 'wrong.' Dogs do understand what is safe and what is dangerous, and they know that it's 'safe' to get into the trash when no one is around to see them …

ANIMALS, WHOM WE HAVE MADE OUR SLAVES, WE DO NOT LIKE TO CONSIDER OUR EQUAL
Charles Darwin

We must remember the human filter through which we view the world, and do the best we can to see with our hearts instead. In this respect it seems that humans and canines are poles apart. Centuries ago, humans elevated themselves high above animals in the belief that these were not conscious beings. Amazingly, this concept is still studied today. While any pet-loving person does not question the notion that animals are sentient, it's good news that we've expanded studies of animal consciousness to include other creatures such as livestock. We've broadened the consideration of animal consciousness and their capacity to suffer and to feel empathy for the suffering of others. If we understand that animals are sentient beings, we can give them their due respect and treat them with compassion.

If we assume that humans are the standard by which to compare other beings, we will miss a lot in understanding other species with whom we share this planet. Ironically, the more we explore how we are different, the more we discover our sameness. Dogs and humans are more alike than not. Still, let's remember that dogs are dogs and humans are humans. Humans exist in anthropocentrism – the sense that everything revolves around us – but we are not the measure of all things, no matter how much we'd like to think we are. Therefore, let's ask the broader question of who dogs *are*, and not just how they are like us.

There is no doubt that the dog is man's best friend. In Chapter 4, we'll explore how wolves domesticated themselves, and, if we observe them in the wild, we can see great similarities between wolves and humans. No wonder they evolved into our companions and confidants as the dogs we know today. We are two species woven together by our social sameness: wolves and humans are socially similar in our pursuit of status. We are alike in the context of how we raise families: in both species, the young are actively cared for by both the male and female in the family, and parents raise offspring over several years to maturity. Like us, they protect the family's long-term survival with the males supporting female and young year-round.

We are kindred creatures with all life, but particularly with dogs because of our sociability and similarity of our emotional lives. Research in the area of animals' emotional and cognitive abilities is expanding, and shows what has been always obvious to those of us who live with dogs: they are sentient beings with feelings and cognitive abilities comparable to our own.

ARE YOU THINKING WHAT I'M THINKING?

By exploring how canines and humans are alike, we become curious as to how each species understands the world. We wonder what dogs are thinking, and if they wonder what *we* are thinking. We question whether dogs think of others as independent beings with their own minds and thoughts. Theory of mind is the capacity to attribute mental states such as beliefs, intents, desires, knowledge, etc, to oneself and others, and to understand that others have beliefs, desires, intentions, and perspectives that are different from one's own. More simply stated, it's the ability to think about what others are thinking.

Without theory of mind and the ability to make inferences about motivation based on behavior, it would be impossible to understand the reason behind others' behavior, and the consequences thereof. Do dogs posses theory of mind? For instance, we've all seen dogs pretend when they play. This ability demonstrates that they understand how others perceive their behavior because the other dog understands that this is pretend. We see dogs signal their intent to play with specific body language (see Chapter 2). Dogs inviting play or making a request demonstrates minds understanding other minds.

My dog, Jesse, was the great pretender. In chapter 1, I relate how she would fool her sister, Sadie, into vacating the bed. Here's further anecdotal proof that Jesse was aware of how her actions were perceived by others (this act could have been nominated for an Oscar). As the story goes, my friend had come by with her German Shepherd, Yogi. My dog, Sadie, and Yogi were pestering my friend for attention as Jesse stood by watching. She clearly wanted in on the action, but was smaller in stature and not the type to physically force herself into the scene. Instead, Jesse used her mental powers. She went to another room, retrieved a tennis ball, and then with Yogi and Sadie watching, tossed the ball across the room. Yogi and Sadie promptly gave chase and Jesse moved in to get attention from my friend.

ANIMAL CONSCIOUSNESS

Defining animal consciousness is beset with challenges because animals do not communicate in human language. The denial of animal consciousness is taken by many to imply that an animal does not feel, which, in turn, devalues their lives. 17th century philosopher René Descartes (I think; therefore I am) postulated that only humans were conscious.

In exploring some of the scientific research, let's keep this fact in mind: we can bring dogs as well as other domesticated or wild animals into the laboratory. But, the real stories of how wild animals behave, and what they understand only happen in their natural environment. Testing animals in contrived circumstances doesn't allow them to show us their true selves, nor the reality of the natural environment in which they live. We have a certain advantage in studying dogs because they would act more naturally in testing. But here, too, we sometimes want to step out of the lab to observe the Lab in *his* world, living *his* life.

DO DOGS UNDERSTAND WORDS?

Dogs understand our behavior and its inference to our intentions. But do they actually understand our words? Enter Rico. Rico was a Border Collie studied by animal psychologist Juliane Kaminski and colleagues from the Max Planck Institute for Evolutionary Anthropology, after reports that the dog understood more than 200 words. Kaminski reported that the claims were justified because Rico correctly retrieved an average of 37 out of 40 items by name.

Another Border Collie named Chaser has shown in tests that she can identify by name 1022 toys and retrieve them. Chaser has shown that she can learn new words by 'inferential reasoning by exclusion.' In other words, she learns by inferring the name of a new object by excluding objects whose names she's already learned.

We might investigate this matter further by comparing canines to kids in their ability to learn words. Developmental psychologists would claim that children also understand the imagery associated with words. They understand the link between an object and a replica or two-dimensional image of that object.

In their book *The Genius of Dogs*, Brian Hare and Vanessa Woods describe an interesting experiment. Juliane Kaminski, who originally studied Rico, designed a test to determine whether dogs had the ability to associate names of objects with their corresponding visual representation. She placed toys in another room, and asked Rico and other dogs to retrieve each toy. Kaminski did not ask for the toys by name as before, however, but instead showed the dogs a replica of the toy to be fetched. All of the dogs correctly retrieved the toys simply by seeing the replica. Studies show that at least some dogs understand the symbolic nature of human communicative signs.

Experiments indicate that dogs understand both what we say and how we say it. A Yale University study shows that dogs can distinguish good advice from bad advice. The experiment involved showing dogs a puzzle box that contained a treat. Researchers demonstrated to the dogs that, in order to obtain the treat, they must move a lever to remove the box top. The dogs learned, though, that it was not necessary to move the lever in order to remove the top, and began to bypass it. (Interestingly, in a related study, young children continued to use the lever, even after it was evident that it was unnecessary to do so.)

Research using fMRI (Functional Magnetic Resonance Imaging) reveals a great deal about how dogs think. The study was about how dogs process human speech; in particular whether dogs understand conflicting messages that they hear. For instance, we might pet our dogs, tell them how beautiful they are, but maybe not so smart. Essentially, we're criticizing and praising simultaneously. The study was done with dogs trained to lie still in an fMRI brain scanner (humane experiment). The dogs listened to a prerecorded series of words spoken by the familiar voices of their trainers. The words were a mingling of praise and neutral phrases spoken with a mixture of positive and negative intonations. Sometimes, the intonation did not match the meaning of the words.

Researchers discovered that, like humans, the left hemisphere of the canine brain is used to process words and the right hemisphere to process intonation, and then combined to understand what was being said. Reward (pleasure) centers of the dogs' brains only lit up when both the intonation and the meaning of the words indicated praise.

Do you hear what I hear?

One way emotions are transmitted across species is by sound, and its subtle tones and tempo. Dogs understand that humans are arguing, even though they don't understand the words. We understand a dog's growl as a warning, and her whine communicates that she is in need or in pain. Quick, ascending sounds generate excitement, whereas long, descending sounds invoke calm. Think about how we make a clicking sound or say, 'Go, go, go!' to a dog to get him to move. When we want the dog to stop, we might say, 'Eeeeeasy.'

Research by Dr Patricia B McConnell, PhD, addressed how the structure of sound influences how an animal responds. In her studies, she recorded acoustic signals from over 110 animal handlers speaking a variety of languages to encourage dogs and horses to speed up, slow down, and go right or left. The results consistently showed that short, rapidly repeated sounds prompted animals to move or speed up, and long, extended, even tones calmed or slowed them.

Dogs modify the sound of their bark, and can even recognize individuals by their vocalizations. They are able to vary timing, pitch, and amplitude. We can certainly notice the difference between an alert bark that says, 'Stranger danger!' and the bark that says, 'Hey! Let me in.' We often think our dogs are barking at absolutely nothing. I jokingly ask my German Shepherd, 'What are you barking at? The voices in your head?' but the truth is that she is barking at something ... it's just something beyond my range of hearing or sight.

Dogs use varying barks and growls to communicate different things. There is a difference in the growl that says, 'Stay away from my food' and the one that playfully says, 'Try and take this toy from me.' We see how the former is a warning and the latter is an invitation to play. Other dogs understand this as well.

Do dogs understand whether or not we can hear them?
Dogs not only hide from our sight when engaging in a behavior they've been scolded for (like chewing on your shoes), they also know when to be quiet, as in when stealing something from the counter.

In their book The Genius of Dogs, Brian Hare and Vanessa Woods describe an experiment offering evidence that dogs understand whether or not we can hear them. Two boxes containing food were placed on the floor. Each box had a string of bells attached, through which the dogs had push to obtain the food. The key to the experiment was that one box's bells had no ringers. Dogs were first allowed to familiarize themselves with the boxes, and then food was placed inside and the dogs were forbidden to take it. The experimenter watched the dogs, and when they did, the dogs tried to take food from both boxes. When the watcher turned their back, however, the dogs took food from the silent box only. It appears that dogs are aware of their audience and knew that the watcher could no longer see them; neither could he hear them as the bells did not ring.

What do I know?
Oftentimes, dogs realise that they do not know how to solve a problem, and many will look to us for help with finding a lost toy, or if a desired object is unattainable, such as food behind a locked door. They stop trying to get it by themselves and ask for help. Some dogs will even bring their food bowl to us when they want to be fed.

What do animals see in the mirror?
There is limited research about what dogs can reflect upon. Here again, if we

use ourselves as examples, we could be missing the point. When dogs are first exposed to a mirror, they might bark a warning, invite the 'other dog' to play, or look behind the mirror to find the 'other dog.' After a time, they lose interest in the reflection, unlike great apes who use the mirror to inspect parts of their bodies that they can't normally see, such as their teeth or backs.

Some suggest that dogs do not understand a mirror image, and this means that they do not have a sense of self-awareness. Self-awareness, simply stated, is the understanding that we are an individual, separate and distinguishable from others and the rest of the world. The definitiuon of self-awareness on a deeper level is the conscious knowledge of our own character, feelings, motives, and desires. By this definition, how can we say that not understanding their reflection in a mirror means that dogs lack self-awareness? It seems all that's proven is that dogs don't understand reflection – or don't care.

This doesn't mean that dogs don't gain a sense of self via other senses, of course. It appears that they are able to distinguish themselves from other dogs through scent, discriminating the scent of their own urine from others. In order to function in the world, all beings must have a sense of themselves, of others, of space, of timing, and how to use their environment to survive. We simply do not know their level of self-reflection.

Do you see what I see?

Canine communication is largely deliberate, whether it be between canines or canines and humans, or any other species. Studies like those of Alexandra Horowitz, described below, explore how dogs adjust their signaling depending on whether or not their audience will receive said signal. Olfactory signals (scent) are forms of communication used by many species, including insects. Scent is a principle form of communication for dogs: odors are long-lasting, unavoidable, and easily placed in many locations. Unlike visual signs, olfactory signals do not require an understanding of the intended audience because the communication is eventually received over time (as in scent marking, for example).

Visual signals happen in the moment, and can only be received if the recipient is watching. Evidence shows that dogs can and do adjust their communication based on what others can and cannot see. Any of us whose dog plays fetch has experienced our dog bringing a toy and dropping it at our feet in front of us. Have you ever been distracted and ignored your dog during this game? If you had your back to her, she most likely picked up the toy and dropped it again in front of you where you could see it. She has demonstrated that she's in tune with what her audience can and cannot see.

What's the point?

Do dogs understand communicative intentions? Can dogs understand that when we point at something, we want them to focus their attention on what we are pointing at? Experiments reveal that dogs can find hidden food when a human points at it. They also show that dogs can do the same by following a subtler cue – the human gaze. In other words, dogs can find hidden food by simply observing where we are looking.

We've explored how dogs understand our intentional verbal communication and their vocal behavior. Now, let's have a look at how dogs communicate visually. We've seen how dogs adjust to their audio audience; let's see how they adjust their visual cues according to how their viewers will receive them.

Visual signals are often subtler than audio or olfactory cues. They are personal and subjective. Audio and olfactory cues are delivered to a broader audience, with olfactory signals designed to be received after the event. Visual signals happen in the moment, and dogs must understand that the receiver is paying attention. To communicate visually, dogs must understand what their audience can and cannot see.

Alexandra Horowitz, author of *Inside of a Dog: What Dogs See, Smell, and Know*, conducted a study on how dogs communicate with each other. She videoed many hours of dogs interacting at a dog park, with her focus on dogs who were trying to initiate play. Horowitz observed that dogs were using visual signals such as play bows only when the other dog could see them. If the other dog was not looking at them, the dog wishing to initiate play would use a tactile prompt, such as touching the other dog with a paw.

In other experiments, dogs have shown that they are sensitive to where our attention is focused. Tests reveal that dogs will beg for food from a person who is facing them or has their eyes open as opposed to someone with their eyes covered or their back turned. Dogs know that we can see them if they can see our eyes.

It is a two-way conversation with dogs. Science is proving that canine communication is a more sophisticated conversation than we suspected. Dogs attune their vocal and visual cues based on the probability of the audience receiving that cue. A 2017 study from the University of Portsmouth found that dogs use more facial expressions when humans are looking at them. They observed that dogs do not change their facial expressions in other exciting situations, such as seeing food. Indications are that dogs use facial expressions to communicate with humans, and know that they are watching. Dogs are sensitive to whether or not their audience is paying attention, and they know how to 'work' us. Who has not fallen victim to those 'sad, puppy-dog eyes?' Yeah, me too!

Do you know that I'm watching?

Dogs seem to be psychic in their ability to read and predict us ... and it seems magical. In my opinion, their skill in this area *is* magical. We are mystified about the level at which dogs can tune in to both us and their world. As we'll discover in Chapter 5, we're all psychic in our ability to tune in at this higher level. Most humans must learn this skill, whereas dogs come by it naturally.

Babies and young children are tuned in to the energy that connects us at this higher degree, but, as we age, we become out of tune and out of touch as the energy is drowned out by the background noise of adult life. Dogs don't tune out like we do because they don't worry about the stock market or when that report is due on the boss' desk. Adult humans forget what it's like to observe, pay attention, and be curious.

Are dogs magnanimous?

In his book, *Beyond Words, What Animals Think and Feel*, ecologist and conservationist Carl Safina describes the wolf known as 'Twenty-one.' Twenty-one was a pup in the first litter born in Yellowstone National Park in the United States, and was part of the program to reintroduce wolves there. Safina spent a good deal of time observing Twenty-one's pack with Rick McIntyre, biological technician for the Yellowstone Wolf Project.

Twenty-one became the alpha male of the Druid pack after the previous leader was illegally shot and killed. He was generous and caring with his pack in that he allowed others to eat first from his kill. He was outstanding in his gentleness, illustrated by his playfulness with pups. During play, he would pretend to lose wrestling matches.

For the most part, life within an individual wolf pack is peaceful. Inter-pack conflict occurs in territorial defense and the expansion of territory. Sometimes fatal conflicts do occur. Among other things that made Twenty-one unique was that he never lost a fight, and never killed a defeated opponent. What could be the reason for Twenty-one's behavior of letting his rivals go free? If a human behaved this way, we'd call them magnanimous. If they released their defeated adversary, their status would be elevated in the eyes of those who witnessed the fight. The loser still lost, but the victor has shown his might and exceptional self-confidence, which are valued traits.

In the natural world, a public show of surplus is sometimes referred to as the 'handicap principle.' The handicap principle is a hypothesis originally proposed in 1975 by Israeli evolutionary biologist and Professor in the Department of Zoology at Tel Aviv University Amotz Zahavi. It purports that the display of excess proves that one possesses enough of something of value, such as bravery or beauty. For example, peacocks display their extraordinarily beautiful tail feathers to attract a mate. In this courtship display, the male bird elevates and spreads his tail feathers, then vibrates them to make a rustling noise.

The shrike, a carnivorous passerine bird, kills his prey such as mice, and then often stores uneaten kills by impaling them on a thorn for later consumption. The speared casualties also serve as advertisement to female shrikes of the male's hunting prowess. During mating season the body count is higher.

Displays of excess could be wealth. In nature, the shrike displays his store of food. In humans, we call displays of an excess of material things or money 'conspicuous consumption.' This is the act of spending money on luxury goods or collecting expensive material things to demonstrate economic power. For example, some people are motivated to live in extraordinarily luxurious homes instead of more modest abodes. Or a person might collect expensive classic automobiles. People can also seek status by taking excessive risks, such as doing extreme sports, or even making high-risk business ventures. In other words, these folks are trying to impress onlookers.

Amotz Zahavi first observed the handicap principle while studying Arabian babbler birds. He saw that the birds would compete with one another for the chance to fight competitors. He writes, "The altruistic act can be considered to be an investment (handicap) in the claim for social prestige, demonstrating the reliability of the claim." When humans behave this way, we could say that we're putting our money where our mouth is. We're not just making a claim about our abilities, we're demonstrating them.

In discussions about wolves and dogs, we often bandy about the words 'alpha' and 'dominance' to describe leadership status. But this language infers that a struggle has taken place in which the leader fought his way to the top position. Modern studies tell a different story. Perhaps Twenty-one's behavior shows a level of altruism. Freeing one's rival demonstrates outstanding self-reliance and strength, which elevates status, and shows that an individual has what it takes to be a leader whom others want to follow, be that leader human, wolf, or bird.

DOGS READ US

Students of our behavior, dogs' ability to observe is seemingly a superpower. They watch us with levels of curiosity and attentiveness similar to that of young children. The problem for humans is that we outgrow that curiosity and attentiveness, and are taught that it's not socially acceptable to stare at others who are different: a human social constraint that dogs don't need to follow. Dogs don't stop observing. They can attune to and observe us at a level that seems mystical; even psychic. Can they read our minds or are they simply wizards at reading our body language?

A story from the turn of the 20th century about a horse called Clever Hans prompted researchers to take a closer look at animal cognition. Clever Hans, it seemed, could count – at least that's what his owner claimed. Hans was shown a math problem written on a chalkboard, after which he would tap out the sum with his hoof. Even though Hans was reinforced for tapping the answers, the questions were not predetermined. Hans was accurate with new problems, and got the answers right even when tested by someone other than his owner. Animal trainers, scholars and the public were baffled. Was Hans actually able to compute the answers to the equations? Was he able to read the tester's mind?

Finally, German comparative biologist and psychologist Oskar Pfungst solved the mystery. He discovered that when the person questioning Hans did not know the answer to the problem, Hans didn't know it either. It turns out that Hans wasn't actually doing arithmetic; nor was he psychic. He was cleverly reading the questioner's body language of unintentional small body movements such as leaning forward, relaxing their shoulders, or small changes in facial expressions when the correct number was tapped out. Hans determined that when he saw these subtle changes in body language he should stop tapping.

Hans is a reminder that animals demonstrate a remarkable ability to pay attention. Our human attention was divided between the math problem and the horse and trainer's behavior. Like a magic trick, we were distracted by the higher concepts of numbers, and were looking for a complex answer. Hans, with no knowledge of mathematics, was able to focus on minute changes in the body language of the questioner. This was the true magic.

Dogs read and 'predict' us by using a combination of their keen sensory skills and finely-tuned ability to pay attention. Thus, they know us intimately, and know things about us that we may not be aware of ourselves.

We've all experienced how our dogs can predict an imminent walk. Of course, putting on your regular walking shoes and taking the leash off its hook cues dogs to the ensuing event, and dogs certainly learn to understand the word 'walk'; so much so that many of us have resorted to spelling out the word so as not to set off the chaotic canine happy dance. (However, spelling is futile because dogs can learn to recognise the spelled-out word.)

Dogs know context. If you were to casually say the word 'walk' while you're doing yoga, your dog knows that you're unlikely to suddenly stop your exercise routine to take a walk. And besides, you're barefoot.

Letting in the dog

But, what if you're sitting at your computer, suddenly look up, straighten your posture and just think about a walk? (I just did that with my dog, Jett, lying on the floor beside me. He responded by getting up and doing his signature invitation-to-do-something play bow.) Dogs, sensitive to our gaze and posture, know that any change could mean a game of fetch might break out momentarily just because we have changed our focus. These movements can be subtle and small, but the professional watcher sees all.

Because dogs read us and anticipate our actions with such prowess, it's easy to attribute character judgment to their skills. I've certainly deferred to my dog's judgment in this matter, especially my people-loving dog, Mr MoJo. On the rare occasion when he reacted negatively to an approaching stranger, I heeded his opinion. Some people even concede to their dog's judgment when choosing a romantic partner. How much of this is actually the dog reading and assessing the other person, and how much is the dog reading our own reaction? Are we being vulnerable to 'confirmation bias?'

Confirmation bias is what psychologists call the tendency to interpret new evidence as confirmation of one's existing beliefs or theories. In this way, dogs can amplify our own beliefs. Let's say that I've 'picked up negative vibes,' from an approaching stranger, and my dog growls at him. I determine that this must be a bad person, because my dog says so with his growl. But, remember how well my dog reads my body language. If I feel afraid, he perceives the olfactory changes that occur with anxiety. He reads the slightest of changes in my body, such as muscles tensing and altered breathing. My dog has also read the stranger's subtle body language, such as direct, prolonged eye contact or darting eyes. My dog has observed the stranger's assertive posture and purposeful stride. We both notice these things and assess the situation. There is a lot of energy being exchanged here where it's easy to blur the line between a dog's ability to judge character and his expertise in observation.

DOGS READ EACH OTHER/SOCIAL INTERACTIONS (PLAY)

Play is important in the social, mental and physical development of dogs. Do dogs play fair? In order for play to be successful, and in order for it to be sustained, it must be just. Dogs maintain fairness during play by doing role reversals, self-handicapping, taking breaks, and signaling intent (frequent play bows to maintain playful intent.) Role reversal means that sometimes the dog is the chaser and sometimes he's being chased. Self-handicapping could be a high-ranking dog displaying a submissive behavior during play such as rolling over. (This behavior would never be done during conflict.) Self-handicapping means that a larger dog will be gentler with a smaller dog than he would be with a dog his own size. Rolling over can be role reversal and also serve as self-handicapping.

Tempered behavior facilitates play and serves to maintain fairness. In order to do this, dogs must be able to read each other's signals and speak their language fluently. If either dog engaged in play deems the encounter unfair, it will end. This will be indicated through visual and audial signals such as snapping or growling. If the warning is not heeded, conflict could ensue.

THE ONLY THING THAT SEPARATES US FROM THE ANIMALS IS OUR ABILITY TO ACCESSORIZE
Clairee Belcher, Steel Magnolias

OUR SAMENESS IS OUR ONENESS

Positive reinforcement works for both species

Positive reinforcement is the giving of a reward following a desired behavior. The reward reinforces said behavior and behavior that is reinforced will be repeated. In training and all relationships, clear communication is paramount. Without it we would be lost.

In training, I set dogs up for success, then use positive reinforcement to let the dog know that he's been successful. Here, we allow learning through making choices. We are building trust by opening hearts, minds, and lines of communication. Training a dog does not mean barking out commands and demanding obedience. That really isn't fun for either of us. In fact, I prefer to think of our experience as teaching instead of training, as the word 'teach' connotes a happier, more interesting encounter. The word 'training' evokes images of boot camp. If we teach with positivity, we can anticipate that positivity is learned and appreciated at both ends of the leash.

Choice and communication work for both species

When people first bring a new dog into their home, they sometimes expect him to acclimatise immediately. Just like us, though, dogs need time to adjust to a new environment and a new life. Imagine that you awoke one day to find yourself on a different planet where you didn't understand the language or the culture. You would be confused to say the least, and probably more than a little frightened. Now, envision a compassionate teacher helping you to learn the language. If a dog cannot understand what we're saying, we need to speak to him in a way that he can understand.

Communication and understanding are empowering; choice is, too. Listening to a dog and giving him a choice is the most compassionate thing you can do for him, although he must first understand that he *has* a choice before he can make it. Too often, we want to dictate and dominate our dogs, and don't allow them choice: by giving them choice, we are choosing compassion.

Dominance works for neither species

The topic of dominance in dogs is controversial among dog trainers, researchers, and the general dog-owning public. Arguments range from the belief that dominance in dogs is non-existent to it being the root of all undesirable canine behavior. But when we erroneously diagnose a behavior problem, we will never be able to resolve it, or even understand it.

The premise that all behavior stems from a dog's desire to dominate lends itself to the use of punishment. In some training philosophies domination implies conflict, which would have us punish the dog instead of helping him through the issue. On the other hand, we shouldn't deny that dogs display dominance because they do. We just don't want to use the concept of dominance as an excuse to punish them.

We might want to explain away unwanted behavior as the dog being dominant, and this is where we run into trouble. As mentioned earlier, dominance is a state, not a trait. Dominance means to employ the most influence in a social interaction. It is a relationship between individuals, and it's contextual. There's what's called situational dominance: for example, a low-status individual might maintain possession of a resource like food over a higher-ranking individual. In a human paradigm, the company president may prevail over her employees at work, but those same employees make the rules with their children at home. In both human families and wolf packs, the parents are always the parents, and their rank is not normally challenged.

Dominance is not winning at the expense of another. It's more nuanced and complicated than that. If we define dominance as winning through aggression, and attempt to train dogs using dominance, then we've made an excuse for abuse. Teaching a dog should not involve conflict. A battleground is no environment for learning. Our goal is to make *both* lives better, not to make one better by making the other's worse. Equanimity brings peace.

Let me suggest a thought about a possible factor in the dominance theory's roots: our attitude about dominating dogs might stem from our propensity to dominate the conversation. Humans are very often talking to dogs and not listening. Maybe we're so worried that our dogs want to dominate us that we want to get the jump on them. (Ha! dominance theory proponents would say that dogs jump on *us* because they want to dominate us.)

Good leaders are in charge of keeping the peace. Instead of using force or intimidation to establish status, they use their heads. In wolf packs, the alpha wannabes are the troublemakers jockeying for position, and the same goes for humans. Corporate middle management, with its dramatic office politics, is commonly where the most discord exists.

Sound leadership is important for successful behavior modification and effective teaching. Whether they're fearful or overconfident, dogs will respond positively to solid, fair leadership over ambiguity. A dog must have confidence and trust in his teacher, and a leader who is above conflict and who sets reasonable boundaries will instill a sense of self-assurance and order in her followers.

When I was in the corporate world, I felt most bad when I thought that I was trapped … when I believed I did not have a choice … when I felt dominated and oppressed. The more trapped I felt, the more desperate I became. This manifested as defiant behavior. Sometimes I didn't show up. I disconnected. The boss who was always reprimanding, scolding, and dominating me lost my participation along with my respect. The same will happen with dogs if we are despot leaders.

Labels work for neither species

We often label dogs as naughty, stubborn, defiant, or lazy – the list goes on – but labels don't serve dogs any better than they serve humans. Labels would have us descend into opinion, and labels lead to expectations. If I treat my dog as a naughty dog because I've labeled him as such, that's what I'm going to get – a naughty dog. It's the simple law of attraction at work. In other words, I get what I expect. The energy I put out into the Universe is the energy I get in return. We need to drop the labels, along with the expectations. Words matter.

As Suzanne Clothier, dog trainer, lecturer, and author, puts it, "Being realistic about what a dog can do and cannot do is an act of love." If I connect to another with loving kindness and compassion, I will, in turn, receive the same. Compassion is the greatest gift one can give to others. It's an even greater gift to give to ourselves.

Unnecessary conflict works for neither species

Let's step back to the topics of dominance and dictatorship. Aggression begets aggression. Sure, I can make my dog get off the couch by physically and forcefully removing him. Some dogs will acquiesce. Others, like my German Shepherd, Jude, would throw me an attitude and show me his teeth, accompanied by a growl. This is unacceptable and unbecoming behavior for a beloved pet, although is a natural way for a dog to express his displeasure, or his response to feeling threatened.

I might have countered his display of dental armaments with the use of aggression. I could have, perhaps, used corporal punishment and forced him to vacate the couch. This would obviously be a foolish move, as the conflict could have escalated (I prefer to spare my flesh from the bite force of a German Shepherd.) Instead, I taught Jude to leave the couch voluntarily, and I did it with positive reinforcement and love. I chose to be above the conflict and, from then on, all I needed to say was, 'Get off the couch,' and Jude would happily move along, as if to say, "Okay. I'm a good boy!"

Make kind choices. We owe it to our dogs to communicate with them in the context of their world. We are also duty bound to help them communicate in the context of the human world where we ask them to reside. Dogs openly share their hearts. We must openly share ours if we expect to understand them in their world, and for us both to live happily and harmoniously in ours.

Visit Hubble and Hattie on the web:

www.hubbleandhattie.com • www.hubbleandhattie.blogspot.co.uk • Details of all books
• Special offers • Newsletter • New book news

43

Mirrors, metaphors
and messages

THE HORSE IS A MIRROR TO YOUR SOUL. SOMETIMES YOU MIGHT NOT LIKE WHAT YOU SEE.
SOMETIMES YOU WILL
Buck Brannaman

I believe that if we are open to the notion and notice, we will see that, oftentimes, our dogs are mirrors. In their non-verbal communication, they can show us where to look, but not necessarily what to see. If we look a little deeper, we'll see the reflection of our souls in theirs. In this chapter, we'll explore how dogs can reflect our moods and our emotions if we are open to their magic. If we look, they will show us metaphors for events happening in our lives. It's like they are little Guardian Angels showing us the way, and reflecting light upon our path. All we need to do is meet them on the trail.

I was walking with a friend and his dogs, and, as was common for us, we were discussing current events. I'd mentioned that prior to our meeting I'd watched a television newscast that had left me unsettled. We were lamenting the whole affair in question, and our attention became momentarily distracted from the dogs. I was pulled back to reality when I looked down to see my dog rolling in poo. Horrified, I quickly removed her from the foul-smelling scene and cleaned her up. My friend shook his head, chuckled, and reflected on the episode and its symbolism. He said that, metaphorically, my dog was showing me how my obsession with current affairs and watching too much TV news was like rolling in poo. It doesn't help anything, and it only makes me feel stinky.

We reflect what we project

When we begin to appreciate the magic of dogs, we can learn to appreciate the magic in ourselves. Dogs express themselves fully, and can teach us how to be the fullest expression of ourselves. They teach us many things if we are available to learn from them. They can teach us peace and that the world need not be a violent place. Dogs can teach us about respect and nobility. They can teach us empathy, and help us to bridge the empathy gap that exists between humans and non-humans. For the most part, dogs are peacekeepers with a desire to live peacefully in this world as a great many of us want to do.

It seems that dogs mirror their human companion's personality traits. People are likely drawn to a dog's personality type for the same reasons they are drawn to their human friends. We are attracted to each other because of our similarities. Calm, relaxed people tend to have calm, relaxed dogs. This phenomenon has not been widely studied, although researchers have discovered, by measuring levels of the stress hormone cortisol, that dogs can take on the personality traits of their people: dogs living with anxious people exhibit anxiety, too, for example. We may indeed reflect each other more than we know.

If you've ever spent time with a negative person, you'll know how easy it is to unintentionally adopt their negative attitude, and this can also happen with dogs. I've seen how therapy and service animals can assume the energy of the people they assist, and negative energy can manifest as physical or emotional illness in the animal, which must be cleared by energy work, exercise, and time off. Service animals and therapy animals can be worked and overworked without regard for their emotional health. This is something that has always bothered me, but I never could quite articulate why; research for this book and studying animal communication has helped me to understand how much animals can take on our energy. If we are attuned to this energy exchange, we can avoid negative effects and strive to keep a healthy balance.

Perhaps we influence our dogs more than we know, but which comes first: the chicken or the egg? When I think of how dogs mirror us, I fondly remember working with a family who needed help with their Golden Retriever. Impulse control was high on the list of things we wanted to address with the high-spirited pup, focussing, in particular, on the dog's overly excited behavior when people entered the home (Dad especially wanted a less exuberant reception upon arriving home from work). During one of our sessions, dad did indeed return home from work, and his lively, 7-year-old daughter raced to the door with the dog at her heels. Both daughter and dog jumped up and down excitedly as dad approached the door. Who was feeding off whose energy? Who was mirroring who?

In Chapter 5 we explore animal emotion in more detail, but, in the context of mirroring, I'd like you to think about how dogs and humans mirror each other's emotions. As described above, the little girl and her dog experienced a shared exuberance when Daddy came home. We can feel a dog's emotions directly, even if we don't fully comprehend his behavior. This is not solely the projection of our own emotions or anthropomorphizing.

The fascinating study of 'mirror neurons' may offer a scientific explanation of how feelings are shared between us. Mirror neurons are neurons that fire both when an individual acts and when he or she observes the same action performed by another. The neuron 'mirrors' the behavior of the other individual, as though the observer was carrying out the act. Studies have yet to reveal which species possess mirror neurons, and to what degree various species can share emotions. Science may prove mirror neurons to be the neurobiological foundation of empathy. In the meantime, let's revel in the reflection of our shared emotions.

GOING WITH THE FLOW
To fully appreciate the magic of life with dogs, we need to let go and go with the flow.

I am a hot air balloon pilot, and the wonders of balloon flight have taught me many valuable life lessons. There is no steering mechanism in a balloon, so you must ascend or descend through the air currents to find one that will take you where you want to go. Talk about going with the flow!

Each time I launch, my landing site is unknown. In flight, I need to stay present, taking whatever winds I can find. If I can't find a wind to take me where I want to go, I accept the fact that I have found a wind to take me where I will go, and maybe even where I need to go.

Strong winds can be frightening, but it has always been the strongest winds that took me the farthest on my journey. This is quite a metaphor for the art of happiness in life. The biggest challenges always teach the greatest lessons. To learn them, one must remain present, be accepting and open. One must let go and go with the flow.

Animals, especially man's best friend, lead us on an intuitive path. Dogs chose to share their lives with us when they domesticated themselves.[1] They chose to join us in our evolution. To me, that's profound, and should be appreciated and respected. It's sometimes difficult to follow our heart's path. We tend to listen to our egos and to other peoples' opinions of what is the correct path in life. Our best friends, our dogs, are our heart's guides. Perhaps they *are* our hearts ... let's follow them.

THE MOST 'DIFFICULT' DOGS
In balloon flight, I accept the winds I find. With dogs, I've accepted the dogs I've found (or who found me.) I believe that we don't necessarily get the dogs we want, but we do get the ones we need, though sometimes the ones we need are

[1]Accepted theory of wolf domestication speculates that interactions between wolves and man became more frequent as humans became less nomadic. When humans settled into villages, opportunistic wolves scavenged human trash. Natural selection began with the most intrepid wolves (probably juveniles who had not yet developed an adult wariness of new things) and humans habituating to each other.

It makes no sense that one of the first animals humans befriended was a predator. However, the wolf's social adaptability and ability to form new social groups (subordinate juvenile wolves separate from their original pack to form new ones) made this possible. The advantages of forming an association with man are obvious for the wolf scavengers. But what's in it for humans? Well, for one thing, companionship, the main reason we choose to live with dogs today.

'difficult.' However, these are the dogs who teach us the most; they are the ones who push our buttons and often push our limits, and it is then that it's easy to become frustrated, thinking that the dog is being difficult, when the reality is that the dog is not being difficult, but he is having a difficult time.

When we are having a tough time, it's helpful to remember that our dogs are going through the tough time with us. Difficulties can manifest in various behavioral changes for all of us, and here we must separate out our own feelings to get a clear view. If we don't, we're looking at the dog's behavior through the filter of our own perceptions. Instead, let's look at the situation with curiosity and compassion. What does the dog's behavior convey about the current situation, or a more general life experience?

THE MAGIC AND THE MESSAGE

Animals are seemingly more connected with the Universe (God, spirit, source, the divine – whatever term you prefer) than are humans. They are connected to one spirit. Animals are concerned with day-to-day survival and do not go to war over principles or politics. Animals can help us connect to one spirit if we look through a broader lens to see their magic and their message. In doing this, we shift our energy to within ourselves, where we find our connection to the Universe.

OBSERVE

Sometimes the lessons we can learn from dogs are less direct; sometimes they communicate by example. As in communicating with the Universe, we must be open to signs. It's possible to observe the relationship between our pets and how we might apply it to our own lives. If you live in a multi-dog household, watch the interactions between them. Some days they are more playful, and maybe they are telling *us* to lighten up. Sometimes they are quiet and prefer to snooze the day away ... would a nap or meditation be in order for me? My dogs like to start the day by racing around the yard. While I prefer to begin my day more quietly, I might at least observe and appreciate my dogs' sheer joy in the arrival of the new day.

MESSENGERS

Sometimes the message is more direct. My dog, Jesse, may have saved my friends and me from the silent killer carbon monoxide. Back in the day when I was regularly flying in hot air balloon rallies, I'd registered for one of my favorite events in South Park, Colorado. My ground crew and I were staying in my crew chief's mountain cabin. The cabin was rustic, but equipped with basic amenities, including a refrigerator powered by propane. The group consisted of three couples, Jesse, and me. Two couples occupied the cabin loft, my crew chief and his wife bunked in their camper, and I bedded down on the couch on the cabin's ground level, with Jesse on the floor beside me.

That night, Jesse was uncharacteristically restless. She even whined periodically throughout the night, and I became puzzled, and even a bit annoyed. The next morning, when my crew chief, Steve, came into the cabin to wake everyone, he was met by what he described as a "wall of dead air." He suspected something was terribly wrong and began to rouse everyone. I awoke feeling groggy and nauseated. Steve realized that there was a propane leak and the cabin was filled with carbon monoxide. Jesse and I got outside while he went to the loft to evacuate everyone there. He had difficulty waking them, especially Audrey, who needed assistance with walking.

Of course I cancelled the flight that day, and we all went into the nearest town for breakfast and to fuel the vehicles. We ran into some EMTs at the gas station and told them our story. They told us that we had been extremely lucky, and had probably come very close to losing Audrey. In hindsight, I can see that Jesse had tried to warn us, and I can also see how I could have paid more attention to what she was telling me. If I had, it wouldn't have been such a close call. Lesson learned.

REFLECTION

Every one of my dogs has been in part, a mirror image of me; their predominant traits a reflection of mine at that time in my life. Some of those traits have remained with me and some have taken their leave.

MoJo was prideful and independent. Bob had a silly side and a goofy sense of humor. Jude did not realize his own power and oftentimes lacked self-confidence. Sadie was optimistic and had a lighthearted approach to life. Abbey was anxiety-ridden and protected herself by lashing out in self-defense. Gypsy was fun-loving, active, athletic, and a risk-taker. Jesse was a creative problem-solver. Penny is sensitive to people's energy and a caretaker. Jett possesses high awareness, watchfulness, and is protective. I can see different levels of all of these traits in myself. It's as if when looking in the mirror, I see all of my dogs looking back at me.

MIRRORS

Most people have a 'favorite' breed, or a type of dog who they gravitate toward. I love all breeds (which is a good thing as a dog trainer) but I am particularly drawn to German Shepherds. I find it interesting that people tend to choose breeds with similar characteristics and personalities to their own. Sometimes I see a mismatch and the problems that arise from this.

German Shepherds are known for their loyalty, tenacity, obedience, protectiveness, and courage. They are working dogs who help the disabled and assist military and law enforcement. They are guardians. Do I possess any of these same qualities, or are these areas of my life that need work? I suggest we can look at ourselves and at our chosen dogs to see what is reflected back at us.

METAPHORS

Below are stories that demonstrate how our dogs teach us through metaphor. I'm not implying that every single behavior that dogs display is a metaphor, or that we should over-analyze and over-philosophize said behaviors. However, if we pay attention, there are times when the symbolism is clear.

Show me the way

I was embroiled in making an important business decision. I'd been afraid to move forward and take this new direction, and was particularly worried about the financial aspects of the new endeavor. I was contemplating this while walking with my dogs. Recently, my dog, Jett, who leads our walks, had wanted to cross a busy street to explore a new area, and, for some reason, I decided to let him do it that day. In the new territory, I found a penny on the ground. I believe finding stray money is a reminder from the Universe that we live in an abundant world. The moral of the story that day, I believe was this: if your intentions are pure and your heart says the venture is right, then take the new direction. The Universe will provide.

What holds us back

I walk with my dogs virtually every day. Upon returning home from our walk one day and as I unleashed Penny, I noticed Jett standing in the kitchen doorway. I knew he wanted to go to his water bowl for a drink as he always does after our walk, but he stood there unmoving, as still as a statue, and I wondered why. Then I noticed that his hind foot was placed squarely on his leash, which prevented him from going through the doorway to get what he wanted. This happened at a time when I was wrestling with a decision about how best to achieve some personal goals. I knew what I wanted, but I was having trouble moving toward my objective. I saw Jett standing on his leash as a metaphor. He showed me how my fear was the leash that was holding me back. Opportunity was knocking on my door, but fear prevented me from going through the doorway to reach my goal.

Opening doors

My two dogs have different personalities, of course. Penny is more communicative – both intuitively and vocally – than Jett is, and will physically open doors by pushing her nose through a gap, which Jett does not do. Penny shows me that sometimes you have to take action and open the door of opportunity. She shows me that, even though I am usually the quiet type, I have to speak up for myself and say what it is that I want and need. My dogs have a lot to tell me if I am aware and observant.

Lighten up and carry on

In the period around Independence Day, Jett becomes afraid to go outside because of his fear of fireworks. I wanted to help him through his fear, and tried gently leading him outside on the leash, figuring that he might be excited at the prospect of a walk. That didn't work. Then I thought of bringing down the energy level instead of raising his excitement level, so I sang a song. That worked, and it did so because it relaxed both of us by changing our energy. I see this as a metaphor for listening to ego. The ego is that (scary, loud) voice in our heads. It's loud and scary like the sound of fireworks. The sound of the heart is quiet, and listening to it is usually the better choice. The voice of the heart is a happy, gentle song.

Don't look back

Penny used to have a habit of walking behind me on our walks, which caused the leash to tap the back of my legs, which annoyed me. She walked behind me because she was feeling nervous, and this position made her feel more secure. The metaphor: notice what is behind me tapping me to say, 'Let it go' then let go of the fear that's holding me back.

I now visualize Penny being brave and walking in front of me. With this visualization and some positive reinforcement, she has mostly stopped lagging behind, and only occasionally falls back if she becomes afraid. She reminds me that we can revert to old behaviors, and are unable to move forward if we give in to fear. We want to stay in our comfort zone where we feel safe.

Don't go there

I had a client who said that her dog was always happy to go to work with her. The dog would run around excitedly every workday when her mom picked up the car keys. My client loved her job and her dog loved being at her side. However, changes at the workplace changed the atmosphere there from fulfilling and happy to highly stressful and unhappy. Now the woman noticed that each day when she picked up her keys, her dog would run and hide: she no longer wanted to go to that unhappy place. She chose to take a cue from her dog and decided that *she* no longer wanted to go to that unhappy place either. She resigned.

Heal

One day, I felt physically 'off.' I wasn't feeling sick, exactly, but I did have a headache and some general achiness, and wondered if I was coming down with something. It was flu season and many people I knew had been ill with a particularly vicious stomach bug. I thought that I should listen to my body and take it easy. I'm not very good at this and have trouble slowing down, and I never nap. I found myself thinking way down deep inside that if I got sick, I could just lie on the couch and watch TV. That sounded so marvelous! But, why wait for my body to force me into that position by getting sick? I can lie on the couch and watch TV, even nap anytime I choose. Why don't I just skip the getting sick part?

I am an early riser, especially when I'm working on a creative project. I'm typically up between 4 and 5am. My dogs usually wake up when I get out of bed, but sometimes they wake me up. At this time Jett had been coming into the bedroom around 4:30am to gently rouse me from sleep. He usually sleeps at the foot of my bed, switching to the couch sometime during the night. The next morning after I'd been feeling 'off,' I didn't wake up until 5:30. Jett was still sleeping on the bed and I took it as his way of telling me to slow down and get rest in order to ward off the flu bug. I spent most of the day on the couch watching TV and napping.

The next morning, I again slept until 5:30. Jett was asleep on the couch, but hadn't come in to wake me. Again, I took my cue from Jett and spent most of the day on the couch watching TV and napping. By the end of the day, I was feeling like I was getting back to normal. The next morning, Jett came in to wake me at 4:34. This was his way of telling me that I was back on track.

During that episode of dodging the flu bullet, I had friends checking in on me via Facebook. I told them that posting pictures of their dogs would help cheer me up and speed my recovery. I was away from my computer for a while, but at around 10am, it hit me that I was feeling much better. Later, when I checked my Facebook page, I noticed that a friend had posted pictures of her dogs right about 10am, the time I noticed that I was feeling better. Wow! That clearly illustrates the healing power of dogs!

Leave it and let it go

A word of warning here: the following story is well, gross.

I took Penny and The Jett to the vet for their annual heartworm testing, and Penny's annual wellness check-up. During the physical exam, we noticed Penny flinch as the doctor felt her stomach. Concerned about that pain reflex, we decided to do bloodwork and a radiograph. The bloodwork was good, but the radiograph showed Penny's intestines were full of fecal matter. This didn't cause a blockage or even constipation, but her intestines were perhaps not functioning efficiently. I mentioned to the doctor that a couple of weeks earlier I'd noticed small pieces of a plastic grocery bag in Penny's feces. It appeared that she'd ingested bits of the bag, which had somehow gotten into the yard. I noticed the plastic pieces over a period of time and was puzzled. Now, you should know that Penny has a dirty little secret. She's coprophagic (she eats feces. I know. Disgusting).

But, here's the interesting part. Just the day before, while working with a client, I'd experienced a stomach cramp. It was so intense that I had to pause the session for a few minutes. This was a highly unusual experience, and, looking back, I realized that I'd had my own dietary indiscretion the day before. I'd had a chocolate emergency and a salty snack crisis, probably brought on by stress as I'd been dealing with some troublesome events and life worries. Like many people, I tend to have difficulty letting go of life's trials, replaying them over and over in my head. I am possibly the queen of mental reruns.

Moral of the story: Penny showed me that I needed to break the cycle. She had eaten plastic, eliminated it, and eaten it again. She demonstrated that I needed to stop the old cycle of taking on the same issues and then not letting them go. I broke the cycle, too. I cleared away the chocolate and stopped trying to solve problems by eating them. I chose to let the Universe handle them instead.

Talk to me

Sometimes the message comes through loud and clear. I became intensely interested in animal communication while researching the subject for this book, and began studying the art of animal communication by taking a home study course taught by a well-known animal communicator. I recently started working with a homework partner, doing readings with each other's animals. In a practice reading, my homework partner spoke with my dog Mr MoJo who crossed over to spirit a few years ago. Let me explain that Mr MoJo was the smartest, funniest, larger-than-life character. He trained me for the nearly 14 years we were together, always running the show. He was a force of nature.

Before our call, my homework partner texted me to say that her phone wouldn't make outgoing calls, and asked if could I call her, so I did. She did a spot-on reading with Mr MoJo, and, as expected, he was very talkative. The next day I was out on a run, training for an upcoming 10k race. I had my phone in my pocket and my ear buds in. I thought I heard my phone dialing but ignored it ... I was running and didn't want to be bothered. Then, a couple of minutes later, my phone rang. I didn't want to be disturbed, so I thought I'd let it go to voicemail. But, my phone answered the call and I heard, 'Hi, Pat.' It was my homework partner. She said she saw that I'd phoned and was returning my call. I told her that I hadn't called! Her phone log said that I had, so we were both puzzled.

Then, it hit me: MoJo! MoJo had determined that if he wanted to talk to me, he should call my homework partner! I'm convinced that's what happened. My homework partner and I laughed and shook our heads. Remembering that MoJo had told her how he'd always liked to make me laugh, she said she appreciated being in on the joke. And he evidently fixed her phone while he was at it. By the way, that's my mischievous Mr MoJo featured on the cover of this book. Can't you see by that rascally expression on his face how he would enjoy playing such a joke?

BUT KNOW WHEN TO STEP AWAY FROM THE MIRROR

I fully believe that our pets mirror us and present us with metaphors. I also believe that we need to know when it's appropriate to look deeper into the mirror, and when we should instead step back for a reality check.

Serious behavior problems need to be addressed by a qualified professional. As a dog trainer specializing in behavior modification, I'm often called to assist with last-ditch efforts or last-straw situations. Perhaps I'm called into an aggression case where the dog has bitten a human or injured another dog. This is not the time for deep reflection about how the dog became aggressive, or how this might mirror the owner's anger issues. Yes, it helps in formulating a training plan to know if the behavior stems from past abuse, neglect, genetics, trauma, or a combination of these things, but let's first answer to the reality of the situation.

The most important objective is to get things under control and, with proper management, prevent further aggressive incidents. Then we'll start the training. Only after the situation is in hand should we take the time to analyze and reflect on any metaphors. Think about it. If you found someone lying in the street bleeding from a gunshot wound, you would not stop to determine the caliber of the bullet before administering first aid, would you?

In the next chapter, we delve deeper into how dogs teach us to tune in to energy, and explore how listening to what dogs are saying opens doors to understanding. We'll learn how connecting with dogs connects us to ourselves.

Visit Hubble and Hattie on the web:

www.hubbleandhattie.com • www.hubbleandhattie.blogspot.co.uk • Details of all books

• Special offers • Newsletter • New book news

Do touch that dial!
Tune into the energy

MAN IN HIS ARROGANCE THINKS HIMSELF A GREAT WORK, WORTHY OF THE INTERPOSITION OF A DEITY. MORE HUMBLE, AND I BELIEVE TRUER, TO CONSIDER HIM CREATED FROM ANIMALS
Charles Darwin

One of my favorite movies is *This is Spinal Tap*, a 1984 mockumentary about Spinal Tap, one of England's loudest rock bands. There is a scene in the film where band member Nigel shows an interviewer the band's amplifiers. Dials on normal amps go from 1 to 10, but all the dials on Spinal Tap's amps go to 11. When asked why that is, Nigel replies, 'It's one louder, isn't it?'

In communicating intuitively with dogs, we need to be like Nigel and turn the dial on our hearts to 'one louder.' The problem is that our egos usually beat us to it. It's the ego, not the heart or the intuition, that turns up the volume. It's that blasted ego! Whenever I'm confused as to whether the voice in my head is my ego speaking or my heart, I ask which one is louder and which one is negative. The one that is louder and the one that is negative is the voice of the ego. The heart speaks more softly, but is actually a *stronger* (not louder) voice, and it's always positive. Let's set it at '11.'

Dogs speak *from* the heart and they speak *like* the heart. My dog, Penny, is very sensitive to energy, and seldom needs to be vocal in her communication with me ... if I'm listening, that is. On the other hand, when necessary, Penny is adequately vocal about what she wants. She will bark at the back door to tell me that she wants in or out. She does her part by telling me; I must do my part by listening. One night I awoke to hear Penny barking *outside* the back door. In a sleepy fog, I was befuddled. My dogs always sleep inside, but, somehow, I'd gone to bed, not realizing she was still outside. She had to vocalize her needs; she had to turn the dial louder for me to hear her.

I'm not saying that all conversations with dogs are non-verbal, but I do believe that the non-verbal conversations are truer. Like all conversation skills, they need to be learned and practiced, though, so, tune in. Trust yourself. Speak softly and listen loudly. And, don't forget to let in the dog!

We're all psychic. We're all animal communicators, though we may not realize it. More specifically, I should say that we don't remember we are psychic. The secret to summoning our telepathic communication skills is to tune in. Collective consciousness is like a radio station broadcasting over the airwaves. It's always there, but if we don't dial in, we'll never hear the message. So turn up that dial.

In this chapter, we explore animal communication. This is not intended to be a 'how to,' as I am not yet an experienced or professional animal communicator. I am a beginner, so will leave the 'how to' to qualified authorities in the field.

In our discussion of canine body language in Chapter 2, we learnt the physical signals that dogs give to express their feelings or intentions. Additionally, we can intuitively 'see' what the dog is feeling on an emotional level. At these times, we often think we are anthropomorphizing, projecting, or being too subjective. Although sometimes true, we are frequently wrong about this. The most difficult thing for beginners in animal communication is learning to sort out what we're sensing – what's my 'stuff' and what's their 'stuff?'

I would guess that if you're reading this, you don't question whether or not dogs have emotions, but, to allay any lingering doubt, let's explore some of the evidence that science has to offer alongside anecdotal evidence from everyday life with our canine friends. For effective communication, we must understand something of the emotional lives of animals, for that is where we make the deepest connection. Here's where we turn up the dial and tune in to the conversation.

We've come a long way in understanding the emotional lives of animals, especially considering that we once questioned whether or not animals felt even physical pain, let alone emotional pain. As recently as the 1960s, scientists maintained that possessing minds and emotions are attributes reserved for humans. Jane Goodall attests to criticism from the scientific community for her 'lack of scientific method' because she named the chimpanzees she studied in Tanzania: a more scientific approach would have been to assign them numbers. She was handed additional criticism for 'giving' them personalities with minds and emotions, and for using the pronouns 'he' and 'she' instead of 'it' when referring to the animals. Using numbers instead of names and it instead of he or she is a way of distancing oneself from the subjects of experiments, and perhaps opens the door to abuse.

Modern science supports the fact that animals are conscious beings with thoughts and emotions. Charles Darwin emphasized that variations among species are differences in degree instead of kind.

In his book, *The Emotional Lives of Animals*, Marc Bekoff includes this quotation by an unknown source: 'If I assume that animals have subjective feelings of pain, fear, hunger, and the like, and if I am mistaken in doing so, no harm will have been done. But if I assume the contrary, when in fact animals do have such feelings, then I open the way to unlimited cruelties ... Animals must have the benefit of the doubt, if indeed there be any doubt.'

Fortunately, science now validates what anecdotal wisdom, open hearts, and experiences of dog people have known all along: animals have feelings. However, the reductionism of strict scientific study cannot paint a full picture of canine behavior. Behavior does not happen in a vacuum; behavior is not simply a response to an environmental or social stimulus. Therefore, we should, perhaps, broaden our scope to encompass both the research laboratory and the everyday 'laboratory' of life.

For the deepest understanding and communication, I believe empathy is the most important of the emotions that we share with dogs. If we deny that dogs have the ability to share the feelings of another, we have no foundation on which to begin communication. We know that our emotions affect our dogs and our relationship with them. We want to be sending a clear message; not one that is confused or ambivalent.

We can only speak from the vocabulary of human emotional experience. In order to communicate with animals, we need to set aside our egos, get out of our heads and into our hearts and theirs. Dogs (all animals, and all of nature, actually) can help us get back in touch with our humanity. In order to do so, we must let dogs be who they are and let them have their own thoughts and emotions. Understand that, even while living in a human-centric environment, they need the time and space to be in their world on their terms.

We need to abandon the idea that humans are exceptional in the emotional arena. We cannot ignore the feelings of dogs in our relationship with them any more than we can ignore the emotions of the human animals in our lives. Not if we want a successful relationship, that is. In order to connect on an emotional level, we must merge knowledge and compassion. We must open both our hearts *and* our minds.

If dogs possess emotions comparable to ours, then we must treat them as such. We'd do well to treat them as we want to be treated: with empathy and respect. Even though dogs may not experience emotions exactly as we do, theirs are just as important to them as ours are to us. The better we understand our shared emotions with dogs, the better equipped we are to communicate with them.

Emotions are social glue. It's easy to share our lives with dogs because humans and dogs share similar emotions. In his studies, Charles Darwin recognized six universal emotions: anger, happiness, sadness, disgust, fear, and surprise. He stated that these central emotions are what help us function in a complex social world. These shared basic emotions are what bond us. Science supports that we share these emotions with other species. Studies show that humans and animals have similarities in our chemical and neurobiological systems, and these commonalities suggest that we have similar feelings. Highly social animals such as humans, with our need to know what others in our social circle are planning or feeling, show emotion with a greater degree of nuance. Here again, we can point out that dogs and humans have both similarities and subtle differences.

Every living creature's emotions matter, and to deny that is to excuse ourselves from the moral responsibilities we have in their treatment. We know that dogs feel emotion, but do they feel emotions exactly the same as do we? To claim that they do would be saying that I feel the exact same sorrow or joy as another human when, of course, we know every being experiences emotions differently.

We can only guess what another species feels. Humans differ greatly from dogs in that our experience of the world is predominantly visual, whilst dogs mainly experience the world through their noses. We cannot and should not compare and judge another species using humans as a template, therefore, as doing so might have us believe that we are superior when the reality is that we are simply *different*. We can't begin to understand or negotiate our differences if we maintain humancentric arrogance about them.

We've explored how dogs learn, how they understand words, and theory of mind. We've examined the canine-human bond, and how dogs communicate through their body language. We need to know all of this in order to understand them, but deeper communication occurs on the emotional level, as this is where we tune in to emotional energy, and where profound connection occurs. Here, we immerse ourselves in the energy and communicate through emotion instead of the spoken word.

I'm writing this chapter in the month of July: a tough month for my dogs as we have a lot of thunderstorms, and our neighborhood celebrates Independence Day with fireworks. Both of my dogs are terrified of these loud noises; so much so that they are refuse to go outside after, say 6pm. This is frustrating for me because the dogs need to go outside to relieve themselves before bedtime. I'm certainly not angry with them, but it's frustrating to know that they have to hold it all night as a result, and I feel bad for them.

I recently had my homework partner in my animal communication class do a reading with my dog, Jett. One thing she got from him was that he was 'feeling a bit outside of the family.' I interpreted this to be his feelings about my frustration. I'd noticed how Jett had not been coming into the bedroom to wake me up in the morning; how he hesitated to approach his food bowl at mealtime until I'd taken a step away. These are subtle cues; cues *so* subtle that many people would not notice them, but I read them as Jett's way of showing that he was feeling disconnected from me.

The reading helped me to see where my frustration was clouding my empathy, and I told my dogs that it didn't matter whether or not they went outside before bedtime. If they needed to relieve themselves during the night and were afraid to do it outside, then so be it. I decided that I'd rather have them make a mess on the floor, than for me to make a mess of our relationship by becoming detached from them. Thank you, Jett, for setting me straight! Once

I made that mental and emotional switch, Jett opened up to me, waking me in the morning again, and soliciting attention and ear rubs. What a wonderful example of how animal communication can help our relationships, and achieve understanding and empathy. Animals just need to be heard.

Do dogs feel empathy?

Empathy is not among the primary emotions listed by Darwin, but is a secondary emotion, which means that it entails a certain degree of conscious thought. Primary emotions are innate and reflexive to external stimuli. Think fight or flight. Secondary emotions such as jealousy and embarrassment are not reflexive, but processed and contemplated before possibly being acted upon.

Empathy is connection. Empathy, the ability to understand and share the feelings of another, is what some say makes us human. Emotional contagion is the phenomenon of having one's emotions and related behaviors directly trigger similar emotions and behaviors in another. For example, a whole flock of birds flies away instantaneously after one is startled. Tuning in with another's fear means matching their emotional state, or, in another word, being empathetic.

We've all experienced in some form or another the phenomenon of having a dog read our emotions. They join us in our joy. Dogs are always ready to party. Start cheering and dancing and dogs will join in. They join us in our sorrow and comfort us with nuzzling or licking away our tears. Science questions whether this behavior is a show of empathy or simply curiosity about what is going on. Is this emotional contagion in which the dog is responding to the human's emotion without really understanding it? Or is it that when the dog sees their human in distress, they become affected by it, respond to their own feelings, and then try to comfort themselves?

An experiment by psychologists Deborah Custance and Jennifer Mayer from Goldsmiths College in London sheds some light on the question. They modified a technique used to measure empathy in human toddlers, testing 18 dogs in their homes with their owners. The dogs' owners and a stranger were seated a few feet apart, and each took turns to speak or hum in an unusual staccato manner, or pretend to cry. The critical point concerned the crying. Researchers reasoned that if the dog was responding empathetically, he would be focused on the crying person rather than on himself, and would offer them comfort (especially if it was his owner, with whom he had a bond). If the dog was responding purely out of emotional contagion, and feeling distressed as a result, he might seek comfort for himself from the person. They further speculated that, if the stranger was crying and the dog responded due to emotional contagion, he would not seek comfort from the stranger, with whom he had no emotional connection.

The experiment showed that the dogs not only approached and offered comfort to their owners when they cried, but also did the same for the crying stranger. Researchers surmised that if this behavior was motivated by curiosity, unusual behavior like the humming would also have elicited a response from the dogs, but they simply looked at the two people, neither approaching nor offering comfort.

The experiment demonstrates what dog owners already know: dogs feel empathy, and respond by offering comfort and support to others who are feeling distressed or unhappy.

Dogs help us to connect to others with their demonstrations of empathy.

Letting in the dog

They show us that we can have empathy toward animals other than pets; those used in experiments, for example, and those we eat. Dogs help us to see that we can have empathy and compassion toward other species, and our fellow man. When we see others suffering cruel treatment and ask, 'Would you treat your dog like that?' most people become upset. There's a meme that goes around the Internet that says we can handle lots of people getting killed in the movie, but don't let anything happen to the dog. If, in our minds, we substituted dogs for other species suffering inhumane treatment, maybe we'd be a more compassionate and caring world ...

Tune in

Dogs express an array of emotions, and we easily recognize the common body language of the happy wagging tail and the relaxed, happy face. To connect emotionally on a non-verbal level is to discover a profound degree of gratification. It's a truer form of communication, which opens and deepens our relationship. We depend heavily on spoken language to communicate and experience life; so much so that, perhaps, it's become our greatest barrier to understanding the world. Language promotes 'otherness,' whereas relating through pure consciousness reveals our similarities with non-verbal animals.

Animals can feel and read our energy on such a level that they seemingly know what we are feeling even before we do. They are tuned in. They are watching. The first step in intuitive communication with your dog is to spend quality, calm time together. Try just sitting with her, touching (if touch is enjoyable for your dog), and connecting with your pet. If your dog is accepting of massage and caressing, then do this. From here, you'll start to notice how your energy is connecting. Your dog is better able to receive your energy and you hers. From this space is where you're able to receive messages and metaphors; information about your dog's behavior, and your own behavior.

As I've said, dogs feel our energy. I find it interesting how my dogs, Penny and The Jett, often approach me as I sit down to meditate. They are drawn to my calm energy. We usually have a little Zen time together with me stroking their ears before I begin my meditation. On the other side of the coin, Penny is so sensitive to my energy that she will leave the room if she senses tension or anxiety in me. She will exit when she hears the 'hold' music of the cable company coming from my phone because she knows there might be a tense conversation.

Canine emotion is unfiltered and true. As a dog trainer and imperfect perfectionist, I see that dogs and humans are not so much different as we are the same. In general, dogs are probably more at peace than humans, because they are better at living in the moment, and never worry about things like mortgage payments and financial reports. Furthermore, I believe it's quite possible that German Shepherds invented yoga.

What brings me to this conviction is the behavior of two of my German Shepherds, Jude and Penny Lane. A few years ago, I began practicing Qigong, an exercise discipline similar to Tai Chi, using slow, flowing and meditative movements. Since the routine included some yoga postures, I unearthed my yoga mat from storage where it had been since before Jude moved in. As I unfurled it, Jude came running across the room, planted his feet dead center on the mat, and assumed the downward dog posture. He did this each day unfailingly when he heard the mat hit the floor (and I have video to prove it). Since Jude's passing, Penny has filled his indelible paw prints on my yoga mat. She is the one who now leads me daily in the downward dog posture, drawn to the calm energy of yoga.

Another example of how dogs read our energy is the story of Pip, a client's dog. Pip was highly reactive to other dogs. She barked and lunged all of her 12 pounds with such intensity that she nearly pulled her mom's arm out of its socket. After a training session, Pip's mom and I were talking. I was explaining how she could reframe her thoughts about Pip's behavior. Instead of thinking that Pip was embarrassingly misbehaved, she could understand that Pip was afraid and having a difficult time. I explained that maintaining a forgiving and accepting attitude would project calm to Pip. We noticed Pip sitting beside us, seemingly listening intently. At the moment I began to talk about projecting calm, Pip lay down on her side and became completely relaxed, showing that she understood.

Intuitive communication with animals

My original plan was to touch lightly on the subject of animal communication in this book. But, as I researched it, I became fascinated with the subject, and began to take courses offered by expert animal communicators. As I said earlier, I am a beginner and do not profess to be an authority, but it might be beneficial to share some of the things I've learned thus far in the hope of helping you achieve a deeper understanding of your dog.

My original opinion of animal communication may be similar to that of many readers. I was open to the idea, albeit with a certain amount of skepticism. I've heard stories of people who had employed the services of a professional animal communicator and were amazed. They had been doubtful at first, but the communicator told them things about their animal that they couldn't possibly have guessed. They asked, 'How could she have known that?' At that point, they became believers.

Before I started studying animal communication, I'd never had anyone do a reading with any of my animals. I began doing readings with my online study group and, upon having the facts from my readings verified by the animal's owners, I would ask, 'How could I possibly have known that?' At that point, I became a believer.

GENUINE LISTENING REQUIRES THAT YOU WILLINGLY BEAR WITNESS TO WHAT SOMEONE ELSE NEEDS TO SAY WHILST SIMULTANEOUSLY SPARING THEM YOUR OWN SOLUTION, DEFENSE, DISMISSAL, ALTERNATIVE REALITY, REBUTTAL, COUNTERPOINT, COMPARABLE STORY, OR MORE EXTREME EXAMPLE. THIS KIND OF LISTENING IS A VERY 'ACTIVE' PART TO PLAY IN A CONVERSATION. YOU HAVE TO BELIEVE FOR THOSE MOMENTS THAT NONE OF THE THINGS YOU MIGHT SAY COULD POSSIBLY BE AS VALUABLE AS HEARING OUT SOMEONE. YOU MAY NEED TO EMPLOY EVERY OUNCE OF YOUR STRENGTH OF CHARACTER TO ACTUALLY PAY ATTENTION, AND NOT BUTT-IN WITH YOUR OWN BIT. THAT KIND OF ATTENTION PAID TO ANOTHER IS POWERFUL MEDICINE.

Gil Hedley

Respect and listen

To communicate intuitively with animals we must engage in active, respectful listening, which is more than passive hearing, and means making a conscious effort to understand the entire message being communicated. Active listening forges positive and respectful bonds if we listen fully by having our minds and hearts completely open.

Letting in the dog

A good listener is present and attentive. A good listener does not judge either end of the conversation. Good listeners must trust in what is received. The first thing for a new animal communicator to learn is to trust what's coming through, separating the message from the noise, and refraining from inserting one's own agenda.

UNDERSTAND

We need to be in tune with our own energy to be in tune with the energy of others, be they human or non-human. Our thoughts create our reality, affecting those around us, and we all too easily project our reality onto others. You know how you can feel your partner's mood when you walk into the room, and how they can feel your mood as well? We are all connected by this energy. I have a saying about some people or animals: 'She really fills up a room.' It means they have a palpable energy that everyone feels.

We're constantly broadcasting our energy to our dogs and our dogs are always receiving it. Dogs are sending their energy to us, too, and we must be aware in order to receive it. For most of us, the more difficult part of the conversation is the receiving. It's easier to talk than it is to listen; easier to broadcast than to receive.

Dogs readily show us that they are listening and observing. If we're sick, they seem to know, and are available to comfort us. When we are sad, they become protective. When we are worried, they will be there for us, acting extra silly to bring us out of our worried state. This is good news, but I also see how people stressing about their dog's ability to read them. People with anxious dogs often blame themselves for their dog's behavior. They say that if they could only relax, then their dog could relax, too. While this might have some truth to it, telling oneself to relax and expecting it to happen is like telling oneself not to think about zebras. See? You're thinking about zebras ...

Penny used to be so afraid to go out on a walk she couldn't even leave the front porch. It was tempting to force her, to show her that it would be okay. It was tempting to try to talk her out of being afraid. But it doesn't work that way. Instead of talking to her about why she shouldn't be afraid, I needed to let her deal with the situation on her own terms. We worked through her fears with positivity. First, I rewarded her for simply looking toward the front walk; for taking one tiny step in that direction. Instead of dwelling on the fear, we were concentrating on her courage. By focusing on the fear, I would have conveyed to her that I expected her to be afraid. By reinforcing her brave steps, I was telling her that she was courageous, and that's what she gave back.

ENERGY

Most dogs love to hear singing, and the beautiful thing is that they don't judge our talent. They don't care if you sing off-key or not know all the lyrics. They love the positive energy of music. Through this shared energy we can learn the dance steps to each other's song.

Dogs recognise positive energy. One day I was walking my dogs in the park when a stranger came gliding up to us on a bicycle. Now, Penny would usually go into orbit at a stranger's approach, but this person was different, it seemed. Even though he had surprised us, apparently appearing out of nowhere, wearing sunglasses, a helmet, and riding a bicycle, Penny gave only a couple of alert barks, then calmed down immediately when I asked her to. I suspect she sensed the man's positive energy.

Turns out, the man had seen my car with my business signs on it, and asked if we were Peaceful Paws Dog Training. He pointed to a now-calm Penny and said, "I want that." He explained how his dog was highly reactive around other dogs, and he wanted to be able to calm him as I'd just done with Penny. He became a client and then a dear friend, and is one of a few (if not the only) males Penny adores. Incidentally, much later, after we'd become friends, I told him that I had not calmed Penny. but that she had done that herself. She knew his energy and she knew that we needed him in our lives – and that he needed us in his.

Relax

To communicate on an intuitive level with animals, we must be calm and we must be neutral. In my practice readings with my fellow animal communication students, we first do a grounding and clearing meditation, letting go of all tensions, physical difficulties, and busy thoughts. Nothing can pass through a contracted mind or body. In meditation, we strive to be like the open sky: to observe any dark clouds passing through and know that the sky is unaffected by the passing storm. Animals will feel safe when talking to a calm, centered person.

Protected

Everyone needs to feel safe in a conversation. In training and communication, I want to convey to a dog that she is safe; especially the fearful or shy ones. Once a dog feels safe, she is more likely to open up and act naturally. Too often, people interpret their dog's behavior as right or wrong; good or bad. Dogs don't understand the human construct of right and wrong, but they do know safe and dangerous. That's how they survive.

Often, people take their dog's behavior personally, telling me that their dog did something just to 'make them mad.' Dogs simply don't do this. They are merely doing what works for them. They know what is safe and they know what behaviors have a payoff. If we listen without judgment and let our dog know that he is safe, we open the door to understanding, from which place we can be the most helpful.

Gratitude

In the context of intuitive conversation, gratitude is not just a perfunctory 'thank you' but a heartfelt, deep appreciation for allowing us in. With wholehearted gratitude, we strengthen the relationship because we show that we value it. We demonstrate how we appreciate those engaged in the conversation.

Be humble

Humans too often view animals as less significant than ourselves. They are different, yes, but not less important. It is madly egocentric to think that man is the template for all life, the model by which all are valued.

We need another, wiser, and perhaps more mystical concept of animals. Remote from universal nature and living by complicated artifice, man in civilization surveys the creature through the glass of his knowledge, and sees thereby a feather magnified and the whole image in distortion. We patronize them for their incompleteness, for their tragic fate for having taken form so far

BELOW OURSELVES. AND THEREIN DO WE ERR. FOR THE ANIMAL SHALL NOT BE MEASURED BY MAN. IN A WORLD OLDER AND MORE COMPLETE THAN OURS, THEY MOVE FINISHED AND COMPLETE, GIFTED WITH THE EXTENSION OF THE SENSES WE HAVE LOST OR NEVER ATTAINED, LIVING BY VOICES WE SHALL NEVER HEAR. THEY ARE NOT BRETHREN, THEY ARE NOT UNDERLINGS: THEY ARE OTHER NATIONS, CAUGHT WITH OURSELVES IN THE NET OF LIFE AND TIME, FELLOW PRISONERS OF THE SPLENDOUR AND TRAVAIL OF THE EARTH

Henry Beston, The Outermost House: A Year of Life on the Great Beach of Cape Cod

Visit Hubble and Hattie on the web:

www.hubbleandhattie.com • www.hubbleandhattie.blogspot.co.uk • Details of all books

• Special offers • Newsletter • New book news

Open the door;
let in the dog

THE PROCESS OF COMMUNICATION CAN BE BEAUTIFUL, IF WE SEE IT IN TERMS OF SIMPLICITY AND PRECISION. EVERY PAUSE MADE IN THE PROCESS OF SPEAKING BECOMES A KIND OF PUNCTUATION. SPEAK, ALLOW SPACE, SPEAK, ALLOW SPACE. IT DOES NOT HAVE TO BE A FORMAL AND SOLEMN OCCASION, BUT IT IS BEAUTIFUL THAT YOU ARE NOT RUSHING
Chögyam Trungpa

We cannot let in the dog without opening the door to our hearts, and we must open the door wide. Our hearts are big enough if we simply remember to listen.

Several years ago I had some remodeling done in my home. The work included the replacement of my back door, which my dogs use to get to the back yard. I was considering whether or not to have the contractor install a dog door. Since my dogs are so big, I asked if he could just place the hinges at the top of the new door instead of on the side, thus making a giant dog door. I already had a perfectly good door, but I needed to figure out how to make it open wide and always be accessible.

Likewise, the door to our hearts needs to be readily available, eternally open, and open wide. In this chapter, I'll be asking you to take a leap into the intuitive side of communicating with dogs, getting out of your head and into your heart. This will be a bigger leap for some than for others, and I'll be asking you to open your mind to the concept of communicating with dogs telepathically. Remember that animals think in pictures. We also think in pictures and use language to convey images and ideas to others. For example, if I say the word 'banana,' you see the yellow fruit in your mind's eye. Telepathy is communication without use of the spoken word. Dogs are finely attuned to the energy of spirit that connects all life. If we are tuned in to this energy as well, from this space we can communicate for a deeper understanding. In applying the magic, we'll be combining intuitive communication and the scientific theory of training.[1]

MUTUAL RESPECT

The unbridled human ego is a risky thing. I sometimes imagine that, given a nice tiara and brass bracelets, I could be the Wonder Woman of dog training. But, inevitably in my unrestrained fantasy, dogs will befuddle me with some new challenge. I am thankful to them for humbling the heroine that my ego imagines me to be.

Successful relationships require mutual respect. I don't live with dogs because I want to dominate and control them. I value life, and I won't be bogged down in the egotistical quagmire of dominance and power over another being. Dogs are incredible, intelligent creatures worthy of our respect.

When I'm walking with my dogs, it's easy to forget that we have different agendas. For example, I might be interested in burning calories, or deliberating over why my phone won't send email. My dogs are interested in sniffing, marking, and more sniffing. If I get impatient with them for snuffling every blade of grass, I remember that they might in turn be impatient with me as I pause to greet a friend. Sound relationships involve giving and taking, and encompass reciprocal respect.

WE MUST LET OUT THE EGO IN ORDER TO LET IN THE DOG

If we operate strictly from ego, communication will fail, for ego attempts to elevate one's status by lowering the status of another. We cannot expect to foster a healthy dialogue with dogs when the human ego dominates it. We cannot connect with our higher selves, nor can we connect with dogs through the ego. We must communicate on common ground.

Dogs and humans are both social species, and here is the common ground on which we can reach understanding. Perhaps it is human nature to elevate our status and to see ourselves as the superior species, but, it may also be human nature to seek something superior *to* us ... supremacy comes with a lot of responsibility, after all. Humans are complex and we tend to equate complexity with superiority. Perhaps we should let go of that egotistical attitude and simply BE. Much of what is important to humans is said without words. All that is important to dogs is said without words. This is an egoless space where communication in its purest form can happen.

We need to beware of metaphorical projections that our egos tend to conjure up, for projections can lead to misunderstanding. While dogs are

[1] I don't mean to suggest that intuitive communication should replace training. Serious or complex behavior problems need to be addressed by a qualified professional.

a lot like us, they are not us. We must recognize the difference between anthropomorphism, projection, and fact. In fairy tales, stories of wolves are not facts about the animal, but merely our projected fears. If we overlook the truth and the scientific facts, if we deny the true nature of animals as sentient beings with emotions, we ensure that they are misunderstood. On one hand, realizing our sameness with another species can be positive and endearing. On the other hand, it can seem threatening when we recognize in them the things that we dread in ourselves. The things that we fear in ourselves are amplified when we recognize them in other species, especially predators like the big bad wolf.

In recognizing our sameness, we naturally might see others as competition. Like different cultures perceiving other cultures as competing tribes to be contained, some would see dogs as competition for power. There are those who think we must dominate dogs in order to keep them in check. This projection debilitates our respect for dogs, and we begin to treat them as inferior. Those we do not consider equal are easily mistreated.

It is what we know already that often prevents us from learning
Claude Bernard

If it is necessary to modify our dog's behavior, we must be willing to modify our own, and we can start by opening our minds, our hearts, and our energy with empathy. Our knowledge of canine cognition and emotion has advanced, and so too should our understanding of how dogs communicate. As we gain a better understanding of and learn more about how dogs learn, we can adopt further creative and proactive strategies in training and behavior modification.

Dominance

Dominance in dog training is a product of the human ego. As a dog trainer I often hear clients ask, 'If I don't dominate my dog, aren't I letting him win?' I respond to that question with another question: 'Letting him win what?' It seems that somewhere along the line dog training has become a perceived battle of wills: a conflict over who is dominant.

But don't worry. Your dog is not plotting to overthrow your kingdom to become the evil overlord. We do want to establish ourselves as leaders, yes, but not because our dog will seize the throne if we don't. We want to be good leaders for the same reasons we want to be good parents. Effective leadership is about setting limits, teaching good manners, maintaining peace in the household, and educating our charges on how to get along in the world. If we don't set boundaries and teach manners, dogs will be out of control, and will 'run the house' in the same way children will if they've not been taught these things. Dogs don't misbehave because they want to be dominant. They misbehave because they've not been taught how to live in the world using our definition of 'behaving.'

'Alpha' and 'dominance' are buzz words used in dominance-based dog training methods, an approach that attempts to emulate the pack behavior of wolves in a dominance hierarchy. Alpha and dominance describe leadership status, but infer that a struggle has taken place in which the leader fought his or her way to the top position. Dominance-based training methods espouse controlling dogs through the use of force and intimidation — methods that include physical punishment such as scruff shakes and alpha rolls.[2]

This approach suggests that dogs regard humans as pack members, and that we should structure our relationship with them accordingly. It says that we should be dominant and hold a superior position (insert human ego here). This method doesn't make much sense because, while dogs may have descended from wolves, they are *not* wolves.

There exists a great disparity between domestic dogs and wild wolves, so the logic of applying the paradigm of wolf pack behavior to domestic dog training is questionable at best. Much of the information used in dominance-based dog training comes from the study of *captive* wolves, conducted in the 1930s and 1940s by Swiss animal behaviorist Rudolph Schenkel. His study concluded that wolves in a pack fight to gain dominance and the winner achieves alpha status. These observations of captive wolf behavior were then erroneously generalized to wild wolf behavior, and also to domestic dogs.

L David Mech points out that this is not normal wolf behavior. Mech is an American wolf expert, a senior research scientist for the US Department of the Interior, currently with the Department of Interior's US Geological Survey, and an adjunct professor at the University of Minnesota in St Paul. Thanks to Mech and others, we've learned that, in the wild, wolf packs consist of a mated pair and their offspring: a family. Sometimes, other families may group together and, as the offspring mature, they separate from the pack. The breeding pair are the only pack members who remain together long-term. Wolves in captivity live in forced packs within a relatively small area, and do not behave in the same way as wild wolves.

We need not throw the baby out with the bathwater by dismissing pack theory altogether, however; nor should we claim that social status in the canine/human relationship is irrelevant, as it can be an important factor in understanding social behavior in both species. Both humans and dogs are communal animals, and both use social hierarchy to avoid and resolve conflict within structures that allow greater social freedom to the higher-ranking individual.

To better understand and communicate with dogs, I suggest we take a broader look at our mutual history – the history of our association with wolves – and how it evolved into our relationship with domestic dogs. In Chapter 4, we took note of the theory about how wolves domesticated themselves. At this point, I should mention that we often confuse the terms 'taming' and 'domestication.' To explain: taming involves behavioral modification: a wolf cub raised and socialized by humans will become tame; domestication entails genetic modification. In the case of wolves, natural selection likely began with the 'friendliest' wolves interacting with humans, ultimately leading to an inherited predisposition towards us.

DOMESTICATION CHANGED DOGS. IT ALSO CHANGED US
As indicated in the study described below, canine brain chemistry may have been altered by domestication, resulting in the practically unmitigated friendliness of dogs. Domestication also changed the way they look to make them more

[2]A scruff shake means to grab the dog by the back of the neck and stare him down until he submits. An alpha roll means to roll the dog onto his back, stare at him, and hold him there until he shows submission. These techniques are considered dangerous and ineffective. The response described as submission is actually the dog shutting down; the danger lies in the chance that the dog may decide to fight back. This is no way to build a trusting relationship.

appealing to us. And it turns out that animals with genes for friendliness look different to those who are less affable. Genes that produce fear- and aggression-reducing hormones are also responsible for physical characteristics that are pleasing to humans: think floppy ears and shorter faces. Our friendliness toward dogs and theirs toward us seem to have evolved in parallel.

The connection between genes and friendliness was proven in a 1959 experiment in Siberia on the genetic basis of behavior. Decades of experiments were done with two groups of captive foxes. One group was bred randomly, whilst only friendly, less fearful foxes were chosen for the other. Surprisingly, over generations, the appearance of the friendlier foxes began to change, with floppy ears, curly tails, shorter faces, and smaller teeth becoming evident. The experiment showed that domestication is a two-way street. Because friendlier animals are more likeable to us, we are friendlier to them. Ultimately, the mutual friendliness between wolves and humans evolved to give us our best friend: the dog.

However, domestication came at a cost for dogs, as surrendering much of their self-reliance and freedom meant they became dependent on us. We've seen that dogs have emotions, which means they are capable of experiencing grief, anxiety, and depression, amongst other things. Living in a our world can cause stress, resulting in psychological problems not commonly seen in wild animals, and it is here in particular that we are obliged to honor their sacrifice and commit to their physical and mental wellbeing.

DEFINING DOMINANCE, STATUS, SUBMISSION, AND AGGRESSION

The misunderstanding and misuse of certain terms does much damage and disservice to our relationship with dogs. Confusion about the ideas of dominance, status, submission, and aggression has compromised our ability to connect with our canine companions. To establish proper leadership and social order with dogs, we must define the concepts correctly.

By definition, status is the relative standing within the social order. High status individuals influence social interaction among individuals within the ranks, and enjoy a higher level of social freedom.

Dominance within a relationship describes one's influence over another, and is contextual. Dominance means to exercise the most control in an interaction, and it is asserted to control, not to destroy.

Submission means to yield to the authority of another in social relations.

DOMINANCE VS LEADERSHIP IN TRAINING

Labels proliferate unfair bias. To brand a dog as dominant colors our perception of his behavior, and translates approaches such as pushiness or willfulness into a dog's desire for power and control. Many behaviors viewed as dominant are less about struggling for rank than they are about maintaining the social unit. Good leadership that manages resources, and sets fair, consistent rules and limits, will peacefully hold the social structure in place.

One problem with the dominance-based training theory is that so many of the behaviors we are trying to modify, such as excessive barking or failure to come when called, have nothing to do with dominance, but happen because the dog is being reinforced for them in some manner. Simply stated, positive training methods look objectively at the facts of the behavior. Knowing that dogs do what works for them, we ask what the payoff for the behavior is, stop the payoff (the reinforcement), and then train a preferred behavior. We're addressing the problem without conflict.

In training, the dominance label limits creativity by setting a tone of conflict. Dogs and humans have complex relationships, where experience and context determine what behaviors are reinforced for the dog. For example, if harsh, painful training methods have been used to train him to stay off the furniture, he may display aggressive behavior when forced to get off the couch. If his first responses of appeasement or avoidance were overlooked, he may resort to aggression. If aggression works, even for a moment, then the aggressive behavior is reinforced. In this case, the aggression originates as a defensive behavior but turns into an offensive one. Here, if we brand the dog as dominant, we limit training to the confines of conflict and anxiety.

What is often interpreted as dominance toward the owner might actually be a dog's confusion and anxiety due to lack of direction. He challenges his owner because he hasn't been taught boundaries or preferred behaviors. Confusion can cause anxiety, where some dogs will defer and some will react with aggression.

In addition to explaining away undesirable behavior, the dominance theory is often used as justification for physical punishment. Aggression begets aggression: I don't use aggression with my dog, because I don't want him to use it with me, and also because it's cruel and unnecessary. A benevolent leader is not a tyrant, but the one who gets to make the rules: the one who leads. A benevolent leader has the ability to elicit the desired behavior from a dog and maintain it with proper reinforcement. Others wait for undesirable behaviors to occur and then punish the dog. Both methods will probably achieve the desired behavior, but the positive leader augments the dog/owner relationship whilst the leader using punishment can destroy it.

Training dogs with positive methods does not translate to permissiveness. With positive methods, a dog learns that polite behavior gets him what he wants (eg sitting before he receives his food, or greeting people at the door with a polite sit/stay garners the attention of incoming guests). A dog who is trained using positive methods will enthusiastically offer behaviors to discover which ones work for him. A dog who has been continually punished in his quest may become afraid of the trainer, and shut down, which is often mistaken for compliance, when, in reality, the dog has given up. A worst case scenario for punishing a dog is that he may decide to fight back.

Our position as leader is not a one-way street; we still need to communicate with dogs and listen to what they are telling us. No one wants to live in an atmosphere of tyranny, confrontation, and total obedience. A caring environment engenders learning and growth.

Being pack leader is not about conflict but the ability to influence others to behave in a desirable way. Let's not wait for our dogs to make a 'mistake' and then correct them. Let's teach them acceptable, polite behavior and reward them for it. Life is so much happier this way, and everyone wins.

In the quest to bust the myth about dominance I came across the following quotes. They might just get us thinking differently about winning and dominance.

"Those communities that included the greatest number of the most sympathetic members would flourish best and rear the greatest number of offspring" – Charles Darwin, The Descent of Man and Selection in Relation to Sex

In his book, *The Emotional Lives of Animals,* Marc Bekoff says he believes that Darwin may well have been right in hypothesizing that sympathetic creatures have more reproductive success. To quote Bekoff: "I propose that this means we should make another paradigm shift in how we understand animals and ourselves. 'Survival of the fittest' has always been used to refer to the most successful competitor, but in fact cooperation may be of equal or more importance. It is likely that for any species individual survival requires both to some degree, while for social species (as opposed to asocial species), the balance may shift significantly, with the most cooperative individuals most often 'winning' the evolutionary race."

"I believe that at the most fundamental level our nature is compassionate, and that cooperation, not conflict, lies at the heart of the basic principles that govern our human existence ... By living a way of life that expresses our basic goodness, we fulfill our humanity and give our actions dignity, worth, and meaning." – His Holiness the Dalai Lama

LETTING IN THE DOG
Letting in the dog is about listening and seeing. If we listen and look with open hearts and minds, we can speak with open hearts and minds. In Chapter 4, *Mirrors, metaphors, and messages,* we explored how our dogs mirror us. My dog, Penny, opens doors, literally: German Shepherds are known for this skill. I've worked with many who are even able to work doorknobs and unlock locks, and I am drawn to German Shepherds. Now I've written a book about opening doors, hearts, and minds. Coincidence ...?

WHERE INTUITION AND INTELLECT INTERSECT
The most open, compassionate, and effective communication with dogs happens at the juncture where science meets the psychic. The better we understand both the intellectual and the intuitive, the more successful our interactions will be.

The Cambridge English Dictionary defines telepathy as the ability to know what is in someone else's mind, or to communicate with someone mentally, without using words or other physical signals.

The academic definition of animal communication is the process by which one animal provides information that other animals can incorporate into their decision-making. The vehicle for the provision of this information is called a signal.

ANIMAL COMMUNICATORS
Animal communicators connect telepathically and intuitively with animals in a silent dialogue. During communication, transference of pictures, words, and feelings occurs between the communicator and the animal. We all have this ability. Well-known animal communicator Joan Ranquet says that telepathy was our first language. Each of us uses telepathy daily on different levels, but professional animal communicators have honed their skills through study and experience, relearning our first language.

Letting in the dog

It can be more difficult to communicate intuitively with our own animals because we are so close to them; too emotionally attached to effectively work together. This doesn't mean that we can't communicate with them, but it can be helpful to call in a neutral party if we're stuck. That way we don't have to separate our 'stuff' from our animal's 'stuff.' The animal communicator will help you to connect with your animal on a deep level and guide you through the experience.

Animal communicators can assist when a pet is exhibiting a new behavior, experiencing a health challenge, or at the end of life. They also offer support when an animal is experiencing life changes, is lost, or has crossed over. The animal communicator does not analyze behavior issues or diagnose illness, but can point you toward the proper action to take with a trainer or veterinarian, or other type of healer. Animal communication can enhance, deepen, and improve our relationship with our companion animals.

DOG TRAINERS

Wikipedia gives the definition of dog training as the application of behavior analysis, which uses the environmental events of antecedents and consequences to modify the behavior of a dog, either for it to assist in specific activities or undertake particular tasks, or for it to participate effectively in contemporary domestic life. Perhaps this is an accurate definition, but I'm glad I didn't try to fit it onto my business card! More simply stated, dog training is a process of teaching a dog to comply with cues, or to modify behavior through classical or operant conditioning.

I am a Certified Professional Dog Trainer. I hold my credentials with the Certification Council for Professional Dog Trainers. I adhere to a code of ethics and employ only force-free methods. I specialize in behavior modification, but also teach basic manners, and work with clients to resolve common problems.

I should briefly explain the difference between behavior modification and obedience training. I dislike the word 'obedience' as it implies submission by the one being obedient, and a dominating authority in the trainer. I don't use the word 'command' for the same reason, and prefer 'cue' instead.

Behavior modification is the process of changing behavioral patterns, such as altering a reaction to situations, people, animals, objects, etc. Obedience means to comply with an order, request or law, or submission to another's authority. Dog obedience training ranges from teaching dogs to respond to basic cues such as 'sit' or 'down,' to high-level competition. But training is more than regimented cues and compliance. A trusting and strong relationship is the basis for success. Truly successful training comes from the heart as well as the mind, and contributes to the wellbeing of both teacher and student.

A TEACHER WHO IS ATTEMPTING TO TEACH WITHOUT INSPIRING THE PUPIL WITH A DESIRE TO LEARN IS HAMMERING ON COLD IRON
Horace Mann

TEACH INSTEAD OF TRAIN

Instead of teaching our dog, let's allow him to learn: a subtle but very important difference between this concept and that of compulsory, dominance-based obedience training. The best we can do for dogs is to present the opportunity for them to learn, and to be successful instead of putting them through boot camp.

Happy, effective training combines scientific method and mindfulness, intuition and intellect; ethology and ethics. Dog training is a two-way street. We must replace a control or dominance mentality with an attitude of cooperation, as doing so diffuses many frustrations that come with our expectations of complete control. Trust yourself and your dog: when you trust your dog, he will trust you.

Lead the conversation

The greatest leader is one who says the least and listens the most. The best leaders don't actually lead anyone; they guide them. Furthermore, followers retain their dignity and feel they are respected; there is a sense of participation as opposed to domination.

This concept is valid in our relationship with dogs as well as humans, and open, compassionate communication is the key. Effective trainers open the lines of communication by becoming astute observers. A dog's first language is body language, so it's essential to become fluent in this subtle form of canine communication.

Let's begin to hear what dogs are saying by using our eyes, our minds, and our hearts.

Mixing the science of training and intuitive communication

We've explored immutable scientific fact and formula, along with ideas that cannot be learned in the classroom or laboratory. If we fill our heads with facts, we must also make room in our hearts for truth. We can create healthy relationships by mixing the lessons of the laboratory with the lessons of real life. We can blend the messages of experimentation and intuition.

How dogs learn

Dogs come equipped with certain innate knowledge – instincts designed for survival, such as breeding and hunting. Beyond this preordained knowledge, dogs must learn the rest of what they need to know about life. We are their teachers for how to live life in the human world.

Dogs learn from environment and experience, as humans do. At birth, puppies begin to absorb information from their littermates, parents, other dogs, and humans. They can learn by imitating the behavior of other dogs, and they will experiment to find out what behaviors work. Early on, puppies learn to read signals; the body language of other dogs. They learn bite inhibition and consequences to their behavior. Effective training takes all of these things into consideration, and channels them into learning through open lines of communication and understanding.

The more complex the society, the more complex the brains of the individuals within it. You require more brainpower if you need to know things like who your competition is or who your allies are. Social animals need to reason and to plan, to protect and understand. They (we) need highly-developed ways to communicate. We must understand as well as express a wide array of components of language, and we must do it within context. Thus, it takes a well-developed brain to build strong social structures. Language is a window into the mind and emotions of another individual. Whether that language is verbal or otherwise, we must strive to understand, because understanding builds strong bonds and trust.

Letting in the dog

APPLYING THE MAGIC

I can be a better trainer if I can include intuitiveness and visualization in communicating with dogs. Additionally, I want to concentrate on the positive, not the negative. Yes, humans have a propensity to focus on the bad. We tend to simply tell our dogs, 'No!' or 'Stop that.' We seem to constantly tell them what not to do, so, instead, let's give them positive direction on what we want them to do.

For example, here's how I taught Penny to stop her disgusting habit of consuming goose droppings. I'll admit, I *was* telling her, 'No! Don't eat poo!' when she was about to steal some of the forbidden 'treat.' Then I thought again, remembering that animals think in pictures. If I'm thinking and/or saying, 'Don't eat poo.' I'm projecting to Penny the image of eating poo. That's confusing for her.

There was an additional challenge in changing her behavior. Some people might ask their dog to keep her nose off the ground completely. That would solve the problem through management, but wouldn't change the dog's desire to eat poo. Besides, I don't want to deprive my dog of sniffing because she needs that mental stimulation. My solution was to concentrate on the positive. I would reward Penny for sniffing and not eating the poo. I changed my thoughts and words from, 'Don't eat poo,' to, 'Sniff,' when she's sniffing only, and then rewarded her. She then began to look up at me when she heard the word, 'Sniff.' Now, she's being reinforced for sniffing and looking up instead of consuming poo.

Even though I am a beginner in animal communication, I've begun to connect with dogs on a deeper level, which greatly facilitates training and takes it to another level. I do not actually do a reading, as I wouldn't communicate with a dog without the owner's request and permission. As a novice, I'll leave that to the experienced animal communicators. But, I can tap in to my intuitive abilities. The following story illustrates how I've begun to combine my skills in the science of training with my intuitive communication skills.

This is the story of Nova, the nervous Malinois. With training and compassion, Nova's mom had brought her a long way from her fearful state, but she was stuck on one thing. Nova wouldn't allow anyone to enter her home without barking ferociously; she even barked at people she knew. Her behavior was fear-based with maybe a bit of territorialism thrown in. On my first visit, I instructed her mom to have Nova outside where she could see me through the glass doors as I arrived. For everyone's safety, I had mom leash Nova, then bring her inside where I was sitting on the opposite side of the room.

As expected, Nova barked madly at me. I envisioned her enveloped in a protective white light and non-verbally told her that she was safe. She calmed down after a bit, but was not able to approach me in that first session. She was just too afraid.

Her mom told me that later that, after I'd left, Nova sniffed the chair where I'd been sitting, then climbed in, curled up and fell asleep. Mom said that Nova had NEVER gotten into that chair before. This told me that we'd actually had a breakthrough; I had, in fact gained Nova's trust a little. She couldn't connect with me when I was present, but she could connect with my energy afterward. She felt safe. In our next session, Nova was able to approach, sniff me, and even take a treat from my hand. Her mom told me that for the whole week afterward, Nova slept in the chair where I had been sitting. By the third visit, Nova was able to interact with me. My ability to read her body language, the use of my own body language and training skills combined with an intuitive connection helped Nova to trust me and feel safe in my presence.

Making the connection

In addition to being a dog trainer I'm also an artist, and I paint pet portraits on commission. Communicating with animals helps me to be a better artist. As I've always been a pretty intuitive dog trainer, I've also been an intuitive artist. I seem to be able to connect with the animal and capture their spirit on my canvas, even if I've never met them. I've had people hug the canvas, cry, and light up when they receive their animal's portrait.

I guess I've always connected subconsciously, but since studying animal communication, I'm able to do it consciously. Now, before I begin a painting I connect with the subject, and ask them if there is anything they'd like me to know. I once did a portrait of two pit bull dogs, a male and a female. The female dog had passed on, but the male was still with us here on Earth. I asked the female dog what she wanted to tell me. She said she'd like to be painted first and that I should work quickly because she wanted to go to her mom to tell her that she's happy and okay. I then asked the male dog what he wanted to tell me about himself. He said, "I'm a goof!" Finally, I asked the female dog if there was anything else she wanted to say. She said, "He's a goof!" That cracked me up.

By the way, the dogs' mom validated my conversation with the dogs when I delivered the piece. Before knowing about my chat with the animals, mom mentioned that the male dog was such a 'goof.'

It's amazing what we can do when we communicate intuitively. In my training and in my artwork, I'm able to make a keen connection with both animals and their people. This kindred consciousness is powerful magic.

Leave the door open: keep the conversation going and the light on

IF THE DOORS OF PERCEPTION WERE CLEANSED THEN EVERYTHING WOULD APPEAR TO MAN AS IT IS, INFINITE. FOR MAN HAS CLOSED HIMSELF UP, TILL HE SEES ALL THINGS THROUGH NARROW CHINKS OF HIS CAVERN
William Blake

In order for our hearts and minds to remain open, we must see clearly. I once worked a case where a senior dog named Sadie had suddenly refused to use the doggie door that she'd utilized her whole life. I inquired if there had been any trauma surrounding use of the door: had she been hurt, or frightened by something while using it? Sadie's mom knew of nothing scary that had happened. I then asked about any upheavals in their lives or medical issues that might be causing stress for Sadie. Her mom said that Sadie had recently undergone cataract surgery, which had successfully restored Sadie's eyesight to normal, and she could see better than she had in years. Sadie's mom then realized that her dog's refusal to use the doggie door began immediately after the operation.

With this information, I decided to experience things from Sadie's point of view so I got down on her level and looked at the doggie door. The plastic flap on it was semi-transparent, letting in light, but it had a cloudy appearance. It occurred to me that Sadie might be associating the cloudiness of the plastic flap with her previous impaired vision. Looking through it may have been similar to how her world looked when she had cataracts. How frightening might that be?

We propped open the flap for an unobstructed view, after which Sadie was quite willing to go through the door. With a little positive retraining, Sadie let go of her fears and continued to use the door without hesitation. The clouds had lifted.

Sadie's story is a good metaphor for demonstrating how communication can become cloudy when our hearts and minds are closed by fear. In this chapter, we'll explore how to keep the doorways open so that we can see our way clear to walking through without hesitation.

IF WE'RE LISTENING TO A CONVERSATION IN THE HEAD, OUR HEARTS ARE NOT AVAILABLE FOR OURSELVES AND THE WORLD, AND WE NEED HEARTS
Cheri Huber

Communication is always happening between our dogs and us whether or not we realize it. Dogs are always listening, and we must be aware of that fact in order to uphold our end of the conversation. Failing to listen means that we are shutting them out, and keeping them out when we presume we already know what they are going to say. We are listening with a predetermined response to what has not yet been heard, and that is not true listening. In order for real conversation to take place we should make no assumptions.

In his book *Canine Confidential*, Marc Bekoff writes, "We must respect and love dogs for who they are, not for what we want them to be."

We have been taught from childhood not to listen ... not to listen to our spirit and not to trust our intuition. We've learned to listen to the ego, and to accommodate society by conforming to rigid rules and standards. We look outside of ourselves for validation and approval. The ego wants to control the world and it wants to control us. Our desire for control and suppression of the spirit results in feelings of helplessness and victimization. The very same thing can happen for dogs, and I frequently see this in dogs who have been trained with harsh, compulsory methods. They comply with commands, but do so only because they wish to avoid punishment. Sadly, many have lost their spirit in the process.

Often, spirit has been lost on both ends of the leash. If we feel powerless, we might attempt to compensate by manipulating and exerting power over another and suppressing their emotions. Many people choose an easy target in this respect — their dog. This imbalance is unhealthy, even dangerous, especially for dogs with fear aggression. Eventually, suppressed emotions will re-emerge: like a beach ball held under the water, ultimately, it will bob back up to the surface. Like us, if dogs suppress their feelings, their energy becomes blocked, rendering them numb. Suppressing emotion and the behavioral consequence of doing so is unnatural for dogs and unhealthy for anyone. We can unwind this dysfunctional relationship by re-educating, reframing, and rebuilding.

A healthier alternative to harsh training is for us to listen to our intuition and to our dogs. We'll use the intellect to support the intuition and vice versa. We'll talk to our dog and let our dog talk to us. The intellect, the intuition, and the dog are all valuable sources of information and guidance. Let's meet at this intersection and walk the path together.

In learning to follow our intuition, it is important to consistently check in with our feelings, which is something I teach when I work with reactive dogs. I ask people to check in with their dog to keep the conversation and the teamwork active. The connection lends support and positive energy to both canine and human. We pay steady attention to the subtle dialog between our intellect, our intuition, and our dog. We work from a perspective of partnership, not of ownership or dominance.

REFRAME

To keep the light shining, we must keep the door open. A good way to do this is by reframing the relationship picture. For clear communication with our dogs, we must let go of our ego and our assumptions. We must set our intention: that we are here to help. I will ask for things from my dog, but she must know that she is safe. She must know that my intent is pure and honest.

In *The Book of Joy* by His Holiness the Dalai Lama and Archbishop Desmond Tutu with Douglas Abrams, the authors speak of the 8 pillars of joy. They talk of joy as being a byproduct of specific qualities of the mind and heart. The qualities of the mind are perspective, humility, humor, and acceptance. The qualities of the heart are forgiveness, gratitude, compassion, and generosity. I'd like to take the 8 pillars of joy concept and apply it to improving our relationship with man's best friend.

OPEN THE MIND WITH PROPER PERSPECTIVE, HUMILITY, ACCEPTANCE, AND HUMOR

We can open our minds by adjusting our perspective to include more positivity: this is the foundation to any happy relationship and in understanding our dogs. We can see a situation as a challenge and be upset, or we can see it as an opportunity. When I'm working with clients and their dog-reactive dogs, I love the moment when they change their perspective. Instead of saying, 'Oh no! There's a dog!' they say, 'Great! There's a dog! We can practice our new coping skills.' They now see the appearance of a dog as a training opportunity instead of a moment of anxiety and frustration.

Once, during a group class where I was training therapy dogs, tornado warning sirens began to blare. We were at the city recreation center where an art class for kids was in session. The emergency protocol for a tornado warning was to move everyone to the centermost interior area of the building.

At first, I thought about the situation negatively. Crowding everyone into a small area along with the dogs could be a disaster. But, then I changed my perspective and thought of it as a training opportunity. The presence of the dogs distracted the kids and also calmed them, and the dogs actually got to practice real therapy work. Everyone was alert yet relaxed with a positive perspective in a situation that could have been chaos. When we step out of our near-sighted self-interest, we can see a larger perspective and a positive solution. Literally and figuratively speaking, the storm passed.

We can open our minds and level the relationship playing field through humility. We've discussed the dominance theory and how it can be detrimental to our bond with dogs. Let us remember that we are all sentient beings connected and not separate from the Universe. We want the same things – to be happy and safe, and to be free from suffering and pain.

Acceptance opens the mind. Without acceptance, we are in denial and nothing much happens in that state. Nothing changes. Acceptance is not equivalent to resignation, nor does it pardon an unacceptable behavior. If we react without acceptance, we cause ourselves pain. We must first accept something before we can change it.

If we regard our relationship with our dog as difficult; if we see them as 'doing it to make us mad,' we'll never get anywhere. If we label our dog 'difficult,' we'll be stuck in a frustrating situation for both canine and human. Instead of passing judgment and regarding our dog in this way, we can open our minds to see that he is not being difficult, but he is having a difficult time. Acceptance fosters compassion and a broader perspective, from where we can begin to improve the relationship if necessary.

Humor also opens the mind. If we can laugh at ourselves, we reduce stress and change our perspective. Sometimes when the weather isn't cooperating or when schedules are unaccommodating, I say, 'It's only dog training.' This doesn't mean that training is not important or serious, but taking a step away and seeing the humor in a situation changes our outlook and we become less critical of ourselves. Once, while working a serious aggression case, I fell into my client's swimming pool. I could have gotten angry, embarrassed, and upset, but I laughed it off. (Admittedly, I was a *bit* embarrassed.) Laughing always feels better than crying, so I choose laughter.

Flexibility and a willingness to adjust expectations save undue anguish. Wishing that Scooby had a penchant to play fetch or to run agility courses when he clearly does not makes both you and Scooby unhappy. It's like wishing one's child had a passion for neurosurgery, when she really aspires to do stand-up comedy. With a certain level of acceptance, we can reach a more contented place in our relationship with dogs. We must meet them where they are, accept them, love them, and forgive their foibles. They are the first to do the same for us.

OPEN THE HEART WITH COMPASSION, FORGIVENESS, GRATITUDE, AND GENEROSITY

We can open our hearts by applying forgiveness on both ends of the leash. It does no good to dwell on past mistakes, and if we accept the situation, let go of anger and guilt, and then forgive, we can move on. Letting go and forgiving doesn't mean to forget. We can remember our mistakes in order to correct them and move forward, and not repeat them. Dogs seem to be experts at forgiveness, and can teach us volumes on the subject. Forgiveness promotes trust. We might ask our dog to forgive us for what we do not know; we might also ask forgiveness of ourselves for our shortcomings.

Gratitude opens the heart. It's easy to be grateful for the obvious, but, by changing our perspective, we can even be grateful for what's not obvious, or even those things that are apparent problems. As in the therapy dog class example above, with acceptance, I changed my perspective, weathered the storm, and then realized how I could actually be thankful for it.

Gratitude expands in order to bring us more things to be thankful for.

Gratitude allows us to savor life with our dogs. It's not happiness that brings gratitude, but gratitude that brings happiness. It's human nature to see the negative – it's an evolutionary survival mechanism – but gratitude overrides the negativity, allowing us to see the positive in a given situation. Gratitude fosters compassion.

Compassion is an open heart, and is defined as sympathetic pity and concern for the sufferings or misfortunes of others. Scientific studies on the subject reveal that humans are not the only species to feel compassion.

In *The Animal Manifesto: Six Reasons for Expanding Our Compassion Footprint*, Marc Bekoff, PhD, a leading expert on animal emotions, makes a call to action in the humane treatment of animals. He asks us to think about how we affect the quality of life for animals in the decisions we make every day. His work shows us that animals experience a wide range of emotions, including empathy. If we deny that animals have emotions, we are blind to their plight. By recognizing it, we can improve their lives and, in turn, our own and the condition of the world in which we all live.

Compassion is the understanding of the plight and suffering of others. As we must have self-forgiveness, we must have compassion for ourselves for our human foibles and frailties. Here is where we develop compassion: lack of it fosters harsh judgment of both ourselves and others. Open your heart with generosity, and be generous with your attention, your time, and your love. Nothing will make your dog happier. In turn, nothing will make *you* happier. Let's remember that generosity is not about giving in order to get. It's about making someone else happy. Our own happiness is the outgrowth of our generosity. Open your heart with generosity of spirit. Dogs do it and they make us happy with their big-heartedness. They can radiate pure happiness and it's contagious. As with forgiveness, dogs can teach us huge lessons.

Now let's explore some qualities of effective trainers that are vital to the art of happy, effective, mindful training.

MINDFULNESS

An effective dog trainer achieves success through mindfulness. The mindful trainer works in the moment, strives to be non-judgmental, and to keep compassion at the heart of their methodology. A skillful teacher is a good listener, a perceptive observer – and has the ability to think on his or her feet.

I have observed these qualities in top authorities on canine behavior, many of whom are human, and many of them have four legs and fur. If we are willing to be students as well as teachers, we will learn more than we will teach.

TEACHING INSTEAD OF TRAINING

My thesaurus lists the words 'teach' and 'train' as synonyms. In my mind, and in the context of dog training, there is a subtle difference between the two.

To me, training infers regimen and obedience, whilst teaching implies a more participatory experience. I prefer to allow my student to learn by presenting an opportunity for success, and then letting them make choices instead of commanding and demanding compliance. Choices empower the student. If I insist on being dominant, I could disempower them and engender fear. I can show that I am worthy of leadership if I choose to empower my students rather than dominating them.

STAY PRESENT

It's human nature to dwell on the past and worry about the future. We get stuck when focusing on the past, and equally stuck when fretting about the future. While in training it is useful to understand the past and have a forthcoming training plan, the most effective teaching happens in the moment.

Staying present means that we don't jump anxiously into the future to live an event that may never happen. Dogs are proficient at staying present. (Maybe I should take a page from their book next time I'm worried about the state of world affairs.) This is not to say that dogs can't remember or think ahead; they don't doggedly dwell on the past and future, as we humans do. With the right motivation and environment for success, most dogs can readily focus on the present task.

NON-JUDGMENT

It is easy to be hypercritical of our own dogs. We may want to hold them to a higher standard, and place them in the 'perfect dog' mold of expectation. We want our dogs to be as 'good' as the neighbor's dog or the faultless pet dog of our youth. However, to achieve effectual training, the temptations of judgment must yield to acceptance. Let's accept the dog we have at present, complete with shortcomings. Only from this point can we move forward on a training path.

Judging people can come ashamedly easy, yet doing so causes great unhappiness. When I'm caught in the throes of judgment, I find it helpful to reframe my critical thought by changing it from a statement to a question. For instance, I once saw a jogger wearing the oddest footgear. He was running in high, black boots fitted with bowed slats for soles that put a lofty spring in his step. The ridiculous footwear made me think, 'That guy is really weird, and he's going to scare my dog!' Catching myself in a moment of judgment, I considered again, and asked myself if the guy really *was* weird. My conclusion: maybe not. Perhaps he had a disability, and those bionic feet enabled him to run. Or maybe he worked for a research and development company as a test pilot for alternative running gear! By changing my perspective, I did not pass judgment, and did not mistake perception for fact.

Note: Reframing my attitude worked for me. However, my dog did bark her opinion of the man's outlandish footwear. She's a German Shepherd who thinks she's the K9 unit of the fashion police.

Dogs are quite skilled at being non-judgmental and non-prejudiced. They certainly can differentiate between genders, colors, sizes, and shapes of both human and canine species. However, unless influenced by an unpleasant experience or lack of socialization, dogs usually choose the path of equality.

LETTING GO

The human tendency to focus on the negative has us seeing only our dogs' 'wrongdoing.' We often expect dogs to automatically know how to live in the human world, and furthermore expect them to know how to live in our individual world (in my world it's okay to sleep on the bed, but it's not okay at Grandma's house.) Dogs are skillful opportunists with a matchless ability to cohabitate with us. They behave in ways that get them what they want. I suggest we make the most of this canine approach by reinforcing the 'good' behaviors offered. Dogs will repeat behaviors that have been reinforced, so let's catch them doing it right and reward them. This is so much easier and more fun than fixing or correcting the undesirable behaviors.

Allowing desired behaviors that dogs offer brings out their creativity and personality. To avoid chaos at mealtime, a dog could be trained to sit and wait before being given permission to eat. This is one outcome, but we need not fixate on it or be attached to someone else's idea of how mealtime should look. My dog, Mr MoJo always sat and 'waved bye-bye' before I put down his bowl. This was his idea, because the trick had been reinforced in the past. I enjoyed his performance of this cute, polite behavior, so it became part of his mealtime ritual.

Overcoming resistance

Resistance magnifies the negative. Resistance is wily, taking on many forms. We don't have to look very hard to find it in ourselves, but we must look closely to find it in dogs – even the ones we deem stubborn. Yet, the act of labeling is, in itself, a form of resistance. (My dog is too stubborn to learn, so I won't even try.)

For humans, resistance often comes in the form of excuses. We say, 'It's going to be difficult' or 'I don't have the time or the right outfit.' Resistance can present itself as distractions such as, 'I'll get to training the dog as soon as I answer the 652 emails in my inbox.' Perfectionism can be a form of resistance, as well. 'I'll never get Sparky to do a perfect recall, so I won't bother.' Happy, successful training happens when we prioritize our goals, adjust our expectations, and set our minds on the positive. And if we catch resistance lurking, we won't sell Sparky short.

Maintaining curiosity and openness

Curiosity goes hand-in-hand with letting go of expectations. Be willing to be a student. Good training benefits from being inquisitive and open to the possibility that the outcome might look different to what's imagined. If we allow our dog's ingenuity and personality to shine through, he will teach us great lessons. This philosophy may not apply to dog sports and obedience trials, but it certainly makes pet dog training a lot more fun. Curiosity and openness are useful attributes for the trainer, as they help us find creative solutions to even the most critical issues. Without them we might dwell on expectations, failing to see the progress that has been made and the power of our innovation.

Enjoying the journey

The path of dog training success does not go directly from point A to point B, but is a circuitous and sometimes uncertain and adventurous course. If we focus too sharply on the outcome, the negatives, the big things, we'll miss out on the little things. (In hindsight, we often realize that the little things actually *are* the big things.) Sometimes we make mistakes and feel that we've strayed off the path, but if we never walk off the road, we'll never find new ground. I suggest we take it all in our stride and enjoy the journey to happy training with those wacky, wonderful canines.

Words matter

Once again, I'm getting picky about language (after all, I am a writer.) The words and phrases we use are important. The Universe and our dogs are listening. I prefer to say, 'I'm walking with my dogs' as opposed to, 'I'm walking my dogs.' The latter sounds more like a chore; the former sounds like a fun activity with friends.

Here's another example of how words matter. When my dog barks, the first thing I say is, 'Thank you.' This validates that she's doing her job and puts me in a better frame of mind than if I simply bark back with, 'Quiet!'

PLAYTIME

I've seen a meme going around the Internet that says, 'Life is short. Play with your dog,' and it makes me a bit sad to think that we need a reminder to have fun with our dogs. Playtime should not be a chore any more than walking with our dog is something to check off our to-do list. I know a woman who designates Saturday nights to playtime with her pets. It's rowdy time for the dogs, the catnip comes out for the cats (in a separate room from the dogs,) and she pours herself a glass of wine. She declares it to be as much fun as a night out on the town with her human friends.

IN OUR DOGS' EYES IS A LIGHT THAT SHINES. OUR WORK IS TO KEEP THAT LIGHT SHINING
Suzanne Clothier

CHALLENGING TIMES

One morning, I was a bit annoyed with Penny. She'd been antsy and pacing, probably because it was raining and we hadn't gone on our daily walk. I said to her, "Penny, I know you're bored and don't understand why we aren't walking. I don't know what I'm going to do with you." Then, I realized the error of my thinking and said, "I mean, I don't know what to do for you." Then we played a game pf chase with a toy, after which she was fine and stopped pacing. Changing my words changed my attitude and my energy, which helped me to find a solution.

We are oftentimes too easily frustrated with our dogs. If she is afraid, we must remember to take a breath and be compassionate. Remember that dogs can be afraid just like us: do you get frustrated with yourself when you are afraid? Think of your dog as being a 2-year-old child. Would you be frustrated with a frightened child or would you offer her comfort?

Once, after a sleepless night and a 7-mile run the following morning, I was extremely tired as I walked with my dogs. Like many people, I get irritable when I'm tired. My excited dogs – who are normally stellar leash walkers – suddenly decided to go in opposite directions. Simultaneously, each one hit the end of their respective leashes and I was nearly pulled off balance. I snapped a reprimand at them, but then immediately felt a surge of guilt, so I apologized. Jett turned to me with a smile and a soft light in his eyes. He very clearly said to me, "It's okay. I understand that you didn't mean it. I forgive you." We walked on. But, in that look Jett gave me, there was something more ... something much more. I felt it deeply. It was a sense of forgiveness that was timeless and all-encompassing. With that one look from one dog, I felt like I'd been forgiven for any and all of my transgressions with every dog I'd ever known. This is the transcendent power of dogs.

Humans tend to be afraid to feel painful emotions. I suspect dogs do not do this, at least on the level that we do. Dogs know what is safe and not safe in the physical world – that is survival. Humans tend to worry about feeling fear or sadness, and that if we allow ourselves to feel the emotions, we'll be stuck there. Humans tend to label feelings as negative or positive and then avoid the negative ones. It's resistance to feelings that make them painful. We can learn from psychologically healthy dogs who are experts at moving through emotions and letting them go.

Dogs are our teachers if we are open to their lessons. Our dogs are always speaking; we must remember to listen to their teachings. Our dogs are always listening; we must remember to listen to them, for all they really want is to be heard. By paying attention, we can leave the door open.

Visit Hubble and Hattie on the web:

www.hubbleandhattie.com • www.hubbleandhattie.blogspot.co.uk • Details of all books

• Special offers • Newsletter • New book news

Coming attractions and canine lightworkers

I USED TO LOOK AT (MY DOG) SMOKEY AND THINK, "IF YOU WERE A LITTLE SMARTER YOU COULD TELL ME WHAT YOU WERE THINKING," AND HE'D LOOK AT ME LIKE HE WAS SAYING, "IF YOU WERE A LITTLE SMARTER, I WOULDN'T HAVE TO"
Fred Jungclaus

Consider. What we think about is what we become. We tend to take life too seriously too often. I say we lighten up and loosen the leash on both ends.

Perhaps, if we look closely, we'll find that dogs hold the secret to living a happy and peaceful life. They can walk with us as canine lightworkers illuminating the path where both dog and human find peace. Dogs can serve as guides who gently take the lead and show us the way to our true spirits.

We are attracted to dogs because they are so similar to us – evolution has brought us together because like attracts like. it's a fine example of the law of attraction at work: the belief that focusing on positive or negative thoughts results in bringing positive or negative experiences into our lives. Thoughts are energy and that force attracts like forces or energies vibrating at the same frequency. Therefore, we can improve our physical lives and relationships by improving our thoughts, and we can improve our emotions by visualizing a positive outcome. By combining positive thought with positive emotion, we attract positive energy, resulting in a positive experience.

There is an energy that connects all life. Dogs know this energy and they are asking us to connect, so let's answer them but raising our vibration instead of our voices to align with the Universe. Alignment with the Universe comes from a place of innocence and purity, uncluttered by problems. Dogs can guide us here because that is largely where they reside. I'm not saying that dogs are never afraid, don't make negative associations, or have bad memories ... they do. They just don't live from the ego and dwell on the negative as doggedly as we do.

When the troubles of the world burden me I sometimes question my life purpose. How can a dog trainer save the world? After all, that's my job, isn't it? Who else will do it if not me? Here, I must remember that the way to save the world is to save my own world by simply shining my own light. If I connect with my true self, my spirit, I'll raise my vibration and my light will shine more brightly.

RAISE YOUR WORDS, NOT YOUR VOICE. IT IS RAIN THAT GROWS FLOWERS, NOT THUNDER
Rumi

Shining my own light inspires others to shine theirs. Teaching people to connect with their dogs through communication and understanding helps them to connect to their own spirit. When we all shine, we can save the world together. All animals, but particularly man's best friend, can help us to see the light. That light is love.

Dazzling examples of rhetoric do not impress dogs; nor do they appreciate lengthy lectures. Our conversations with them must happen straight from the heart: connecting requires a deep trust in our dogs and in ourselves. Let's give dogs the credit they deserve for their ability to communicate via pure energy, for reminding us that we are all connected in the web of life, and for how they help us grow by being a positive influence for us. Thus, we can become a positive influence for the world.

THINK LIKE A DOG

The law of attraction advocates reframing negative thoughts into positive ones, but needs clear communication to work. If my thought is, 'I'm poor,' I'm telling the Universe, 'I don't have enough money.' This thought comes from scarcity, so the negative thought I'm sending into The Universe is, 'I need more money.' The Universe can only return what it receives, so what I will receive in return is the need for more money. For the law of attraction to work and bring me abundance, I must convert that negative thought into a positive one. Rather than thinking I need more money, I might think, 'I live in an abundant Universe. I will always have enough money.' In this way, I am aligned with positive energy and the Universe will reciprocate with abundance.

Like the Universe, dogs need clear communication to understand us. As we learned in Chapter 4, dogs give back what they receive. If I want my dog to stop jumping on me, I might think, 'Don't jump on me.' But, if I'm thinking about the dog jumping on me, then that's the image I'm projecting to the dog. Therefore, that's what I'm going to get back ... a jumping dog. If I want to change the behavior, then I might think about my dog sitting politely for greetings instead of jumping up. I can change that negative thought to, 'sit, please,' and, with training and positive thoughts, I can get that sit.

In training, we sometimes say, 'Train. Don't complain.' Complaining generates negativity. When speaking with a potential training client, I love to hear a positive attitude – one that says training will make the dog happier instead of the mindset that the dog is misbehaving. Oftentimes people think that their dog behaves badly in order to make their life miserable or to annoy them. This negative thinking makes for negative experiences. Instead, let's strive for a more positive outcome by declaring our intentions, framing them in a positive way, and undertaking positive actions for change.

Let's investigate some constructive approaches to achieving positive results.

Conceive

Thoughts generate the actions we take in our lives: what we think about is what becomes our reality. If we tense up and anticipate negativity, that's what we'll get. If we relax and expect positivity, we'll get more positive results. It's a self-fulfilling prophecy.

It's important in our relationship with dogs to keep our thoughts positive. Instead of focusing on what we don't want (an anxious or reactive dog, say), let's focus on what we do want – a calm dog. (I have a complete treatment plan for reactive dogs. For this discussion, I'm focusing on just one of the tools in my training toolbox.)

We can't teach a dog what not to do (react); we can only teach a dog what to do instead.[1] We can teach our dog an alternative behavior in order to achieve our goal of calm instead of chaos, but, even after doing that, we must maintain positive thoughts. I've seen dogs who have fallen into a pattern of alert, react, retreat and then recover. Both dog and human were anticipating the problem, so continued to attract it. We had to take a few steps back and project positive thoughts at the moment the dog alerted, which allowed us to share their calm, not their chaos.

Positive thoughts become confident actions and confident actions attract positive outcomes. Negative thoughts become feeble actions; then feeble actions attract negative or inadequate outcomes. The law of attraction operates strongly on our subconscious. Because dogs are attuned to us on this intuitive level, they can reflect our subconscious thoughts. Since we're more confident with positive energy, we project positivity, which dogs pick up on and mirror. That's a plus! When working with clients on this concept, I kindly remind them not to beat themselves up for their all-too-human reflexive reactions. I encourage them to have compassion for both ends of the leash, and remind them that this is a process.

Because humans filter images through language, we must remind ourselves that animals think in pictures. Therefore, we need to hone our visual thinking skills for more effective communication with them.

We'd been working with Jasper's reactivity to other dogs for some

---

[1] In treating reactivity, we are not simply training a behavior; we are teaching the dog new coping skills resulting in a calmer, happier dog. For serious and complex behavioral issues, seek the help of a reputable professional.

time. I realized that, during our sessions, his dad and I were both anticipating Jasper's reaction of spinning and barking wildly at the sight of another dog. Our expectations were projecting that image to him, and he was reflecting it back. We worked on changing our thoughts and, consequently, the image we were sending. Instead of the negative thoughts, we began imagining Jasper alerting to the dog, but remaining calm. Then, we'd relieve the pressure by giving him some space from the other dog, which demonstrated our support for him and reinforced the calm behavior.

I felt that part of Jasper's reactive behavior stemmed from his perception of himself as our protector, so we told him that he didn't need to defend us or be in charge. We told him that we were a team and that we had his back. The first time we did this, I noticed that Jasper made a stronger connection with his dad than he had before, and saw a change in Jasper's face and his demeanor because we'd allayed the frustration and anticipation. We didn't take away his 'job,' but helped him with it. Because he's a proud and independent fellow, it was important to let Jasper make his own choices, and we'd empowered him by letting him choose calm. By changing our thoughts and expectations, we began sending a different image; thus, we were able to modify the behavior. The result? A happier, more positive experience for everyone.

Receive

Pay attention. Dogs do. How often do we walk with our dogs and think about things like finances, schedules, or what we're going to make for dinner? Let's remember that dogs experience the world predominantly through their noses, and their agendas are very different from ours. Instead of being impatient when dogs stop to sniff, take the opportunity to look around, listen ... maybe even do a little sniffing yourself by breathing in the air. How does it smell? How does it feel? You might just notice a sign from the Universe.

A few things that people believe to be signs from the Universe are white feathers (a sign that the angels are watching,) a coin (a sign that we live in an abundant Universe), a hawk flying overhead (hawks are said to be messengers of the animal world), or white butterflies (angels are nearby). You may even see an actual 'Stop' sign near a bed of flowers ... maybe it's telling you to stop and smell the roses?

Perceive

If I perceive or expect an event to be negative, it likely will be, and then I will feel it's my duty to fix it. In training dogs, I set them up for success and reward them for it, instead of waiting for them to make a mistake and then correct them. We can train our minds in the same way. If I perceive a situation as negative, my response will be to correct it, but why not look for the positive and set myself up for success? How rewarding!

For example, if my dog tries to eat something on the trail that is bad for her, I can take it as an opportunity to teach her to leave it alone instead of simply correcting her. I can make it a teaching opportunity or I can set the stage for conflict, and fight to get her to drop the object. With positivity, we both win. She gets rewarded for leaving the item and I don't have to take something gross out of her mouth.

How we perceive situations determines how we will experience them. If we anticipate and label something as 'good' or 'bad,' that's the experience we're liable to have. Withholding judgment helps us to stay positive. There is a story that comes from the Taoist tradition: a tale of a farmer whose horse has run away. The farmer's neighbor comes by to express how sorry he is to hear the bad news about the missing horse, and the farmer replies that we can't really know what is good or bad. The neighbor is thoroughly confused by the farmer's attitude.

However, the next day, the horse returns and brings with him 12 wild horses. The neighbor happily congratulates the farmer on his good fortune. Again, the farmer simply states that we can't really know what is good or bad.

The following day, the farmer's son attempts to tame one of the wild horses and is thrown off. The son's leg was broken. The neighbor offers the farmer his condolences at this bad news. Again, the farmer says that we can't really know what is good or bad.

The very next day the army comes through the village recruiting every able-bodied young man to go to battle in the war that was raging in their country. The farmer's son is passed over because of his broken leg.

The moral of the story: we can't really know what is good or bad. Stay present and stay open and curious to what the future will bring.

BELIEVE

If I believe that change can happen, it will. What we give to dogs is what we get back. If we are stressed, confused or afraid, our dogs may mirror these mental states. Both dogs and humans can respond to emotions similarly. When stressed, we might leave the room or walk away. We might look away or feign disinterest by reading a text (a dog would intently sniff the ground) Both species might try to avoid the source of stress by not paying attention to it. We can believe in our dogs and ourselves in order to create change and build confidence, and believing in each other is the beginning. If we are trying to change our dog's behavior, we must be willing to change our own, and believe that we can be that change.

CANINE LIGHTWORKERS: HOW DOGS ILLUMINATE US

I am quite aware that dogs do, in fact, speak, and they are often the voices of reason. They are lightworkers in their own right and we can follow their light. They are guide dogs on the path of enlightenment if we allow them to take the lead.

Lightworkers, by definition, are highly sensitive and receptive to energy, tuning in to the moods and feelings of others. They know how to turn on their own light by following the positive energy that makes them happy and allows them to shine; then they share the light.

As we've seen, dogs excel at reading and following energy. They have the ability to follow their hearts as they follow their noses. If we wish to learn from them, they will teach us how to find our light. Dogs are masters among us who shine their lights brightly by purely being themselves, and we are raised into the light by simply being in their midst ... but only if we watch and listen. Let's be aware of their specialness and how they live in the world; let's not reduce them to mere animals, which would have us seeing our otherness more than our likeness, and thus dimming our lights.

I received this reading with my dog, Jett, in my online animal communication practice group. Jett has a message for human beings: he told me, "Tell people that we can talk like you but in silence (telepathically.) Please listen to us. We are here to help each other build a community and to be happy."

LUMINOUS BEINGS ARE WE, NOT THIS CRUDE MATTER
Yoda

FINDING SPIRIT
Dogs know how to blend the rational and the intuitive, whereas humans seem to have unlearned this skill. Our spirits are lost in the world of form and survival of that form. We revere logic (ego), and too often dismiss instinct (heart). In my opinion, dogs exist among the ranks of spiritual teachers and gurus: true lightworkers if we let them shine. We are used to giving dogs commands, but we would do well to take a few cues from them. They can help us focus on our true nature as lightworkers, for if we look to see their light shining, we will recognize our own inner light.

FINDING AWARENESS
Dogs can bring us back to awareness and remind us that we are living in a magical, wonderful world. They wake us up and can instantly bring us out of our negative thinking. They help us turn from worry and fear to full-on, in-the-moment silliness and joy, and do so by tuning us in to a lighter energy; showing us our hearts by connecting us to theirs. They show us that our minds and hearts are not separate. Our minds try to protect us, as our dogs do. Our hearts know our true purpose, as do our dogs.

Dogs remind us to pay attention to the ever-present positive energy, and to notice how life is filled with moments of enchantment. We are lucky because dogs don't judge us but give us never-ending, unconditional love, teaching us about absolute love by simply giving it freely. As we've discussed in Chapter 1, we've chosen each other to be family by attracting each other's energy. Their purpose here is to show us what unconditional and pure love looks like, and that it exists everywhere. They can help us to see ourselves as they see us. Perfect.

FINDING THE LIGHT
Dogs literally bring us into the light. My dogs are the inspiration and guides who brought me to this place in my life and career, and to writing this book. They've led me along this path to fulfilling my dream of helping people and their dogs. When I received the offer for a publishing contract, one of the first things I did was to go outside and share a happy dance with my dogs, because I knew they were instrumental in bringing my dream to fruition. They showed me how to follow my heart and to fetch my dreams. They showed me unconditional love and how to give it to myself in order to find my life purpose and to shine my light. They show me that the light can shine any time. Always.

REFLECTING THE LIGHT
Dogs are masters at shining and being themselves, and they are masters of authentic communication and giving unconditional love. By showing us love, they show us that we deserve to have love – and not only from them, but from ourselves also. They have open hearts to show us how to open ours to ourselves and to others.

Dogs are expert at being in energetic alignment with the Universe.

Additionally, they are in energetic alignment with us. If we choose happiness, they mirror us in our joy, and are the guard dogs of our joyfulness.

Dogs can be traumatized and show us how they need forgiveness and compassion to heal, just as people do. I do the bulk of my work with fearful and fear-aggressive/reactive dogs: although, I say it's my work, I am really part of a team in which we're helping each other – owner, dog, and coach. In working these cases, I can help people to calm and center themselves in order for their dogs to become calm and centered. I begin to see a cycle as the human watches their dog find composure, and then they, in turn, become calm. The mirroring process works both up and down to both ends of the leash.

I have been drawn to helping dogs with issues (training), and giving them a voice (writing), and doing so can help their people, but I've also helped myself. With the study of animal communication, I'm getting to a deeper level with discovering how much more dogs have to tell us and teach us. This is the world of the unspoken – the world of intuitive communication, which speaks even louder than the spoken word if we listen with open hearts and minds. When we teach, we learn. In teaching, I sometimes learn about canine behavior and sometimes about human behavior, though, if I look closely, I see that I'm usually learning about both. Furthermore, I learn about myself. I'm still discovering the depths of the canine/human bond, in which I cannot help the dog without helping the humans, including myself.

My journey into animal communication has shown me how I can help dogs in ways I didn't know existed. I've always been open-minded and curious about animal communication, albeit this was mixed with more than a touch of skepticism. I only became a full believer when those 'How-could-the-communicator-know-that?' moments began happening to me.

For example, I did a reading for a friend with her Great Dane, Simon. He told me several things about himself, including one curious thing. I got the word 'stars' from him. In our discussion of the reading, Simon's mom was unsure about the meaning of this word, but, later that day, I received a text from her with a photo attached: Simon's mom had realized what 'stars' referred to. The photo was of the tile floor in their home's entryway, in the center of which was a decorative design with a star. His mom said that this was Simon's favorite place to sleep. He liked to have his bed there and if she moved it, Simon would drag it back to be by the star. He liked the spot because the sun shone through the etched glass of the front door, creating warmth and casting refracted light (more stars) on the floor. There is no way I could have known these things. This was one episode that clinched my belief in animal communication.

Dogs help keep us in the 'now.' They live much more in the present than we do, and don't obsess over the future or lament over the past in the same way as us. Getting stuck in the past by feeling guilt, or caught up in the future with worry is such a human thing. Let's allow our dogs to keep us attached to the present.

By shining his unique light, your dog will show you some of life's secrets for happiness. Dogs live a simpler life, whereas humans succumb to unnecessary drama, detail, and discussion. Dogs live their truth – they can help guide us to live ours.

When we live our truth we live in an exhilarating energy. Once, when I was running, I came up behind a 5-month old St Bernard puppy walking on a leash with her people. I always approach dogs carefully when running; reading their body language and energy so as not to startle them. This pup looked back over her shoulder at me, gave me a big smile, then started to pick up the pace. Her people said, 'Hey, it looks like she'd like to run with you.' She had picked up on my happy energy: I was in the zone, connecting with nature, my body, my mind, and this puppy. Yes, in general puppies love to run, but there was something about this pure energy, the pure connection of that moment with that pup. We were both in our truth. In subtle moments like these, dogs can teach us how to find joy in life. If we pay attention, dogs will help us grow into our truth.

ANIMALS ARE HEALERS AS WELL AS TEACHERS

There is no denying the human-animal bond. Whilst we are estranged from many species, we have made the acquaintance of dogs (by their ancestors' design). Amazingly, the science of how dogs think, learn, and communicate is recent. How is it that we've not thoroughly studied our closest friend and confidant before? Have we mistakenly thought that the intangible is the unknowable? Much of our relationship with dogs is intuitive, but is the unprovable incomprehensible?

THE HUMAN-ANIMAL BOND

The American Veterinary Medical Association (AVMA) defines the human-animal bond as "a mutually beneficial and dynamic relationship between people and animals, that is influenced by behaviors that are essential to the health and well-being of both."

The Human Animal Bond Research Institute website states, "There is growing evidence that companion animals positively influence many important physiological, psychological and relational benefits, including quality of life and well-being." The website also states, "HABRI is committed to supporting scientific research to substantiate what many of us know to be true, that humans and pets share a special, mutually-beneficial connection."

Research shows that human-animal interaction enhances our quality of life, reducing stress and depression, getting us moving, bringing value and meaning to our lives, reducing blood pressure, and raising self-esteem, amongst many other beneficial things.

Pet Partners, an organization whose mission is to improve human health and well-being through the human-animal bond, also recognizes the mutual benefits of our dynamic relationship with animals. Scientific research is catching up with what you and I already know: our relationship with animals and especially our pets has a positive impact on our own physical and mental health, as well as the physical, emotional and social well-being of the community.

Studies indicate that people living with companion animals have lower blood pressure, higher survival rates after suffering a heart attack, a reduction in risk of cardiovascular disease, increased positive social behaviors among children with autism spectrum disorder, and a reduction in health care spending due to fewer doctor visits.

Pets get us out of our heads, give us an outlet for nurturing, and are trustworthy confidants. I am co-founder of Professional Therapy Dogs of Colorado, whose mission is to promote the human-animal bond for professionals working or volunteering in the mental health field. PTDC provides educational workshops, training, and evaluation for dog/handler teams. We are dedicated to enhancing the mental health professional's ability to incorporate and deliver individualized animal-assisted therapy (AAT) interventions.

The impact of therapy animals on improving the physical, emotional, and social health of clients is powerful, positively affecting pain levels, and increasing motivation to participate in treatment.

I had a personal experience that illustrates the power of animals to assist us in healing and overcoming fear, which also helped to validate the trust I'm building in my animal communication skills.

I had been involved in a serious car accident – not badly injured, just genuinely bashed up. I did not see any of my clients the following day, but instead dealt with the tasks necessary to get my life back on track, one of which was to secure a rental car. To my surprise, driving the rental car the 3 miles home was terrifying: the accident had traumatized me more than I knew. I was also nervous about seeing one of my clients the following day, as she has two Great Danes (one of which outweighs me). They are truly well-behaved, gentle giants, but I was nervous nonetheless. (They have been known to have outbursts of playful exuberance.)

That night, I dreamed about the dogs. They told me not to be afraid, they would be extra calm and gentle with me. I kept the appointment and the dogs were true to their word. They went extra slow down the stairs, and the one who outweighs me simply spent a few moments leaning against me in the sunshine. We really connected and she helped me to become grounded. It wasn't just a dream. Those dogs really talked to me. Yes, they did.

Connecting with animals opens hearts and doors to enhance the connection with other beings. Animals are pure in their purpose with no agenda or nefarious intent, and they always speak the truth. Animals are so authentic because they live from the heart, and live and love unconditionally. They are role models to teach us absolute love, for they show us the deepest parts of ourselves.

Animals help us to know ourselves. I sometimes push myself too hard with an I-have-something-to-prove attitude. In order to show the world that I am not lazy, I'll push myself to exhaustion. When this happens, I know that I've mistakenly listened to the hard-driving, negative voice of my ego, and I must quiet that voice (I often yell back at it!).

Jett has a way of showing me the importance of relaxation, and I'm learning to follow his lead. Jett is not lazy, but he knows how to do his job and live life without pushing himself. I doubt that he has a voice that he must silence, but here I can learn from Jett as well, who is the opposite of my pushy ego voice. His voice is gentle, and reminds me that I do not need to yell at the pushy voice which is really only trying to help. Would I yell at Jett for trying to help? No, so, I'll stop screaming at my ego's voice, thank it, and then follow the gentle voice of my heart instead. Jett shows me how.

WE ARE BORN AS INNOCENTS. WE ARE POLLUTED BY ADVICE
Henry David Thoreau

TRUST AND TEAMWORK

We are taught from childhood not to trust our feelings, and not to express ourselves fully as our true selves, and we often attempt to impose this attitude on our dogs. We can have apparent success by punishing away the behaviors (ie lunging and barking in fear at the end of the leash) when, in reality, we've only suppressed the behavior. When the dog again becomes afraid enough, she will revert to the old behavior, just as humans would.

People come to me for help with their reactive dogs when they've reached the end of their ropes, and their dogs are literally at the end of their leashes reacting to what triggers them. Things seem to be falling apart, and the people are frustrated and frightened, just like their dogs. Frustration on the human end of the leash can turn to anger, which might turn to punishment, and only make matters worse. Now, both ends of the leash are feeling victimized. There is a positive way out of this destructive spiral, however.

Let's first take a moment to recognize that the dog is upset, and usually frightened also. Punishing her is not helpful because, whilst it might suppress the unwanted behavior, it won't change how she already feels inside (afraid, threatened, and a victim), and now she is also the victim of our punishment brought on by our frustration. We need to be patient with her and ourselves while we relearn how to handle the situation, and not punish ourselves, just as we would not punish a child for falling as they learn to walk.

If we use punishment and aversive equipment such as choke chains or prong collars, we fail to work with our dog as a team, and we fail to work with compassion. Aversive methods create a learning disconnect. Think about reading this book, and trying to learn while someone is choking you. You would be unable to absorb any information, let alone retain it.

Punishment merely *appears* to work. A dog will comply, but only to avoid the punishment. If we wait for the dog to 'disobey' and then correct her, we are operating in reactive mode. Let's see how we can go from reactive to proactive.

The first order of business is to deconstruct the relationship: let's take it apart before it falls apart and go from being reactive victims to having a plan and working as a team. A good training program for reactive dogs provides both human and canine with specific coping skills. Equipped with the proper tools, they become a proactive team instead of reactive victims.

It was our second session with Bailey, the Shih Tzu. We'd just learned how to relieve anxiety and pressure by turning on cue and distancing ourselves from what triggered her (dogs). Suddenly, a dog appeared on the path ahead. Her mom had Bailey turn and put some distance between them and the dog. Up until then, Bailey had been hyper alert and distracted on walks. She and I had never quite connected intuitively; there was an unseen wall between us.

After being surprised by the dog, her mom helped Bailey by giving her what she needed – space from that scary dog. Instead of putting Bailey into a situation that she couldn't handle, her mom showed her that she was there to support her.

Something clicked; Bailey turned to me with a big smile on her face and happily jumped up on me. It was very clear to her mom and me that she was thanking us for understanding her needs. We'd relieved her from the stress and pressure of having to deal with the scary thing on her own, and had empowered her by giving her a choice: thus, she was no longer a helpless victim. We'd also empowered Bailey's mom by giving her and Baily new coping skills. Neither of them were victims any longer. In that moment, trust was built at both ends of the leash.

We cannot live fully and joyfully if we're living in fear. When we learn that we are not victims, and not give away our power to things outside of ourselves, we take control of our lives. When we go from both ends of the leash being scared victims to empowerment of choice, we are free. We do it together as a team.

We don't have to be perfect; we just have to be real. Dogs are masters at being real. Our deepest relationship and the one that matters most is the relationship we have with ourselves. Our outer life and all relationships are merely reflections of that. Dogs are real and they can help us to see our own reality. Old school training methods have strict rules that we expect dogs to adhere to. We are demanding and commanding; talking not listening. We are looking for something from the dog to help us fulfill our ego's desires. Dogs need to know that they can influence their world and us. Dogs will tell their own story if we are willing to listen: they want and need to be heard.

If we listen to the needs of the dog instead of our egos, we facilitate understanding and communication, and these are the pillars of a real and fulfilling relationship. If a dog is not giving us what we want, perhaps we should take a different approach. If she's not learning in the manner that we are teaching, then let's change our teaching style to accommodate her learning style. This illustrates how we can improve our relationship with ourselves by listening to our own needs. By doing so we learn to listen to the needs of others and vice versa. Listening allows an energy exchange to move between us. Dogs have a powerful connection to their spirit, and we can help them operate in the physical world whilst they, in turn, can help us connect with spirit. They will not disappoint us so let us not disappoint them. We can show each other the way.

Animal communication has opened doors of opportunity for me to help dogs in powerful ways, but first I had to find those doors before I could open them. As it turns out, the doors were in my heart. It's true. Enlightenment is not something found outside of us but on the inside. We find it by listening to spirit – our own and the spirit of all life.

Too often we fail to listen to our dogs and to our own soul. Everyone simply needs to be heard. If we listen to dogs, they can show us their amazing ability to follow their spirit with honesty and spontaneity. Let them guide us to our own spirit by listening – really listening – to them, and hearing them.

RECOMMENDED READING

Taking the Lead without Jerking the Leash, The Art of Mindful Dog Training
• Pat Blocker CPDT-KA

Body Language, A Photographic Guide • Brenda Aloff

On Talking Terms with Dogs, Calming Signals • Turid Rugaas

Beyond Words, What Animals Think and Feel • Carl Safina

Canine Confidential • Marc Bekoff

The Genius of Dogs • Brian Hare and Vanessa Woods

Inside of a Dog: What Dogs See, Smell, and Know • Alexandra Horowitz

Plenty in Life is Free: Reflections on Dogs, Training, and Finding Grace
• Kathy Sdao ACAAB

Animal Speak, The Spiritual & Magical Powers of Creatures Great & Small
• Ted Andrews

How To Be Your Own Genie • Radleigh Valentine

The Book of Joy • His Holiness the Dalai Lama and Archbishop Desmond Tutu with
Douglas Abrams

The Emotional Lives of Animals • Marc Bekoff

INDEX

Letting in the dog